I

L

O

V

E

B

E

V

Unified Field Theory Two
More Hidden Secrets

By John Hildreth Atkins
Jonathan G. Rundy

Dedication

This book is dedicated to the thousands of people, of all walks of life, for whom the task of finding a Unified Field theory has brought them to a wall of obscurity.

I would also like to dedicate it to an innocent man, Mr. Frank Edward Gable, and to Mr. Robert Taylor, administrator and owner of the website: www.freefrankgable.com

Table of Contents

Brief Synopsis

Atkins takes up where he left off more than five years ago. Many of his theories, then in their infancy stage, have now progressed to an astonishing conclusion. His COW formula explains why planets circle the sun. Magnetic Reconnection interacting with the Earth's plasma field propels us towards the sun. And he explains why planets rotate on their axis. The end result is that he proves, beyond any doubt, that there is no such thing as gravitational attraction. It just isn't needed. And astrophysics, as we know it today, is absolutely wrong.

Introduction

Though I have been fascinated with the solar system since I was old enough to know that it existed, I never gave it any serious consideration until about ten years ago. It was at that time that I decided to unravel the mysteries it contained. Little did I realize that my discoveries would ultimately lead me to a Unified Field theory.

I think the problem for those who preceded me lay in their mistaken beliefs. That is to say, these people failed in their quest because they were working with flawed science. Naturally, any efforts expended that involved this erroneous science would, itself, be doomed to cataclysmic failure.

Part of the blame for the erroneous science comes on the shoulders of Albert Einstein. Al's genius was not so much in his fanciful musings as much as it was in getting people to dare to look at the cosmos in a different light. Ironically, media reports that only a handful of people understood his work compelled millions of people to jump on the proverbial bandwagon.

Al's brilliance was in creating a fantasy world built on supernatural events, thus replacing conventional science with ghosts and goblins. Life after Einstein engaged us with galaxy gobbling black holes and space warps that mesh time and dimensions in a cosmic hocus pocus. Nice.

I, for one, was not amused. It all seemed absurd to me. That absurdity reached a climax, of sorts, when Physics Review rejected my work because it repudiated Einstein's theories. PR editors hastened to direct me to read their guidelines for submissions. You see, my work would not even be read once they discovered that I disagreed with dear old Al. I was incredulous.

Do we live in the dark ages? How can mortal man seek to censor knowledgeable people solely because we do not subscribe to popular beliefs...even when we can disprove those beliefs? I was outraged; to say the least.

Nonplussed, I continued in my quest to understand the universe. I wanted to know the truth and did not wish to be mired down by misdirection, conventional wisdom (sic), or scholarly restrictions, imposed by 'droids who lacked the ability to reason.

I laugh now. You see, if PR was shocked that I would attack the credibility of such an illustrious man as Albert Einstein, how much more chagrined would they be when I challenged an even more highly revered man? Holy cow.

Reason told me that I could not accept any belief that involved action at a distance. The notion that gravity was a force of attraction was both mind-boggling and mind-numbing. Moreover, it was predicated on the work of Sir Isaac Newton; a man who lived hundreds of years before Einstein. And it was wrong.

Fact is, if I can explain all motion without the need for any such force as attraction, then there is no empirical data anywhere in the known universe with which to repudiate my work. Many will try, we know that for a certainty, but how does one argue with the truth?

Through the years, I strived for perfection. More than that, I worked with two goals in mind. First, I surely did want to know the truth. Secondly, I wanted to receive accolades which professed me to be the man. That is to say, I was desirous of a Nobel Prize for my work.

When I revealed my goals to a man who worked for Homeland Security, he promptly, and without even a whimper of hesitation, informed me that I would never get a Nobel Prize. Fact was, I wanted to be the first man to get one in every category. How disheartening to discover I could never get one!

When I queried as to why this man felt I would never get a Nobel, he replied that it was because I was not Jewish. Hey, if it will get me my coveted award, I am Jewish. Sign me up.

Since that time, I have come to realize that he was quite right. Officially, 25% of all Nobel Prizes have gone to Jewish recipients. Officially. Unofficially, I would say that that number is somewhat higher---closer to 90% with the other ten percent being to sympathizers or as bribes.

Why is it that when it is illegal for a government employee to accept gifts, all of these officials are getting Nobel Prizes for things they did not do? I cite Barack Obama as one. As soon as he was elected President, he was awarded a Nobel Peace Prize. Incredibly, he hadn't done anything to merit one...though later he announced that he was going to use the United States military to defend Israel. Of course he was. That is what bribes are used for.

As I researched what this man had told me, I discovered that the Nobel Prize Committees were located in countries that are, historically, heavily Jewish. That, however, was not where the correlation ended. You see, the Nobel Committee sends out queries to heads of the major Universities to ask them who they think should be nominated for a Nobel. So I looked at that and boy was I shocked.

Most, if not all, of the heads of all of these Universities, of which they contact, are Jewish. It is historical fact that the Jewish people do not like to marry outside of their race and they almost always refuse to employ gentiles...at least not in high positions.

So my friend in Homeland Security was correct; I can never get a Nobel Prize. So long as Jewish Committees are querying Jewish heads of colleges, my name will never appear on any of their lists.

Yes, I was shocked at the revelation, but there was more to come. Two years after the Nobel Committee bribed Obama, they awarded three Jewish men a Nobel Prize in Physics for the notion that the Universe is expanding and the further away from us, the faster it is expanding.

Not only is that idea redundant, it was the brainchild of a man who advanced it in 1923...long before any of these three men were even born. That man's name was Hubble (yes, the very guy who they named the space telescope after). Moreover, it would have been impossible for these three frauds to have measured those distances without using Hubble's law for calculating same.

Perhaps the most humorous thing about that, and believe me, it is humorous, is that the theory is fraudulent, too. I won't explain that right now but will expound on it in a later chapter. In that chapter, I will give you the true reason why the data seems to bear out the idea of an expanding universe. Meanwhile, you should understand the simplest reason why it isn't so.

According to science, no matter which direction they look, objects in space are moving away from faster and faster the further away they are. Problem is, if that were true, then that would make the Earth the center of the Universe. Need I remind you that that has been tried before?

If it is true that Jews are really supporting one another and censoring the rest of us, no wonder science is in the dark ages. That is primitive, irresponsible, and tantamount to stepping back into the dark ages. Small wonder science is in such a quagmire that the best thing they can concoct is a scheme to oust Pluto from the planetary lineup.

By the way, I shall also explain why they were wrong to do so. Actually, they were right to do so...just for the wrong reasons. Moreover, Mercury, Venus, Earth, and Mars, should all be booted out, too. You're going to love that explanation. And you will know it is the truth. However, just like me, by the time you finish reading this book, you are going to wonder why they don't give me a Nobel Prize!

In the Beginning...

The bastions of science have created a boondoggle of sorts. We are bombarded with daily references to Big Bang and the whole Event Horizon thing. And I suppose, if I smoked enough dope, and it was superior quality stuff, I might, maybe, perhaps, possibly, see where a Big Bang was feasible. However, there is not enough dope on the planet to get me that stoned.

The Universe is infinite; it has no beginning and no end. It has always been, and will always be. No, not as we know it now, but infinite just the same.

Let me clarify something right now. Just because I believe that the universe is infinite does not equate to a disbelief in an almighty creator. On the contrary, I am a firm believer in God; though your version will probably be somewhat different than mine.

God is not a deity. He (or she, if your prefer) does not sit on a throne passing judgment. God is an energy form. Let's call it a creative force from which all knowledge passes. For example, have you ever wondered where a thought comes from?

We are all tapped into a Cosmic Consciousness. Every thought we have emanates from that collective body. Seek and thou shalt find. Listen and you will hear. Be quiet the soul.

Each of us tunes into that body according to our efforts and prayers. Now, I do not mean prayers in the sense that one asks God to do something. Those are good, and substantive, but not the only way to the truth, the light, and the way. Let me tell you a story.

Many years ago, I used to ride the city buses a lot. It was a great way to meet people and to share ideas. A story started on one of town in the morning would pop up on the other end of town that evening. Great, if you want to start stories. :)

Anyway, I am an amiable sort who always says good morning to people I encounter. And so it was that I frequently said good morning, or hello, to these two mentally challenged girls who always sat up front. Invariably, neither would reply to, or in any ways acknowledge, my salutations.

One morning, I was riding in a seat about three-quarters of the way towards the rear of the bus. It was a slow day and not many people were riding. Out of the blue, those two mentally challenged girls got on the bus and one of them said hello. Was I imagining it?

I looked up and she was looking right at me. She repeated her hello. After a quick look behind me, and seeing that there was no other passengers on the bus, it was readily apparent that she was talking to me. So I smiled and said good morning to her.

I noticed that, as the girl sat down next to her friend, her friend jabbed her in the ribs. Simultaneously, the two got into a discussion about talking to strangers. They resided in a group home where they were taught never to talk to strangers.

"He's not a stranger," the one girl insisted."

"Yes he is," the other girl said with due diligence.

"No, he's not a stranger," the first gal repeated. "You know him."

"No I don't," came the curt reply. "Who is he?"

"That's Will Smith," the first gal stated with absolute conviction.

To this revelation, the second girl countered, "That's not Will Smith; Will Smith is a black man."

It was a good point. I am an old white guy. Everybody chuckled; the bus driver probably the loudest. Myself, I was quite impressed. You see, I knew the rest of the story, and that validated what the first girl had declared.

No, I am definitely not Will Smith. I am much poorer than he is. But here is the part of the story that I knew.

I had been writing, and publishing, nonfiction books for several years when I decided I wanted to try writing fiction. More specifically, I wanted to try writing a screen play. Having read an article wherein Will Smith stated that he wanted to do more science fiction movies, and he wanted to do only the ones that were believable, I decided to write a screen play for Will Smith, and using my version of science.

My question is, are mentally challenged people really dumber or are they receiving too much information all at once? How would you like to be bombarded with a trillion bytes of data streaming into your head every second of the day?

I swear that is a true story. I did not make any of it up. Makes you wonder. More than that, it confirmed my suspicions that we are all like radio receivers tuned into this vast body of knowledge. Each is tuned to the station we like. Unless, of course, Mother Nature intercedes.

No, I do not believe in a Big Bang. It isn't necessary, and there really isn't any proof of one. In fact, all of science disputes its existence. The much-touted photograph that showed an even heat signature across the visible universe is easily explained by more conventional science.

One thing that I noticed about most, if not all, galaxies, is the fact that they tend to be flat. This tells me that each galaxy has experienced at least one "big bang." No, not the sensationalized Big Bang, but a much smaller version.

I think of a galaxy as a living, breathing, organism. Each expands until it reaches a point wherein space and light from other galaxies causes it to collapse back into itself. This creates a tremendous pressure and pressure, as we all know, generates heat energy.

As you envision that whole process, think about where the greatest force is coming from. By noting the direction of the greatest application of force, one can easily surmise the path of least resistance. This is important because the resulting explosion is going to follow the path of least resistance and that is going to be tangential to the galactic alignment. In other words, the galaxy will explode into being at right angles to where it presently lies.

Incidentally, I believe that all force is tangential in nature. It makes for some very interesting observances.

One thing that never ceases to amaze me is how so many people, with science degrees in hand, can look at a black hole and not see it for what it is. Perhaps it will help them if I just go ahead and state the obvious.

A black hole is merely a place where matter is not. Nothing more; nothing less. And do you know where these awe-inspiring goblins lurk?

Almost all black holes are found in the center of a galaxy. There's your first clue, Sherlock. It is the only clue you really need.

Every get on a Merry-go-round? Ever notice how, as it rotates on its axis, everything gets flung off towards the outside? Well, think about that galaxy as a very large merry-go-round. If we ever see a galaxy that doesn't have a black hole at the center, I shall think something is very wrong.

It is feasible that a newly formed, or reborn, galaxy might not, as yet, have an absence of matter at its center. Other than that, since all galaxies rotate on their axis, inevitably they will all have black holes. Yawn.

"Okay, wise guy," you orate with contemptuous glee, "explain how it is that light gets drawn in at the Event Horizon and debris is sometimes seen shooting out the center of a galaxy?"

Elementary, my dear Watson. Elementary. With an absence of matter, there is nothing for light to reflect off of. Therefore, it appears to be sucked in by an extreme gravitational field.

"And the debris?"

That, too, is easily explained. Have you ever heard of a solar wind? Scientists have measured a solar wind of roughly 509.17 miles per second at the Earth's orbit. That's one hell of a lot of wind.

Now imagine that solar wind is passing through the center of a galaxy at right angles to the galactic alignment. Are you starting to see the big picture?

The Infinite Speed of Light

I purposely left off where I did in the preceding chapter because I was about to encroach on a subject that is better dealt with in a new chapter. The reason I say this is because I am plying you with new ideas. Each new idea should have its own chapter for you to review. Humor me. After all, everybody has their own idiosyncrasies to deal with.

Within a closed system, such as a galaxy, light has a speed limit. This limit exists for two reasons. First, it must interact with matter within the confines of the galaxy. And second, the presence of matter drastically alters space; which, in turn, impedes the progress of light.

I think people have gotten so complacent about light that they no longer think of it as radiated energy. When you look around the universe, there is really only one form of energy that is readily visible. I.E.- radiated energy.

A fact of life is that light energy is the only visible form we see. We know that there are thousands of wavelengths which our eyes do not discern. Many of these are deadly.

Another fact of life is that all forms of energy, on this planet in particular, originated from our sun. All forms. This is probably the most important fact that you will ever remember. It is extremely important. As you shall soon see, it is far more important than you ever imagined.

But let us return to the propagation of light

As I said, light within a closed system is inhibited by the mass of that system. Many of you are now asking the relevant questions: Why do I refer to a galaxy as a closed system? And does that infer that there exists an "open" system wherein light is free to accelerate?

A closed system is any system that contains matter...excluding, of course, the perceivable universe. Absent a sufficient amount of force with which to propel matter, it is quite impossible for matter to enter into interstellar space. This interstellar space contains galaxies which, in turn, allows us to classify them as closed systems.

Since there is no matter in interstellar space, light is free to continue accelerating. For those of you who are hell-bent on addressing dark energy, you can say that interstellar space is made up of dark energy and that enables light to continue accelerating at speeds much higher than 186,231mps.

If that does not satisfy you, feel free to attach a boson (such as a tachyon) to our light packet. However, in reality, I do not believe that any of those things are needed.

Think of that photon as it races across a system such as our solar system. It is going lickety-split from one end to the other. There is a point in the beginning, right before it exits the sun, where light is traveling far less than 186k. Suddenly, like the breaking of a rubber band, it pops out and goes on its merry way.

This light is very much a wave function and the presence of so much material in space hinders its' progress. When it gets to the far end, right where the solar system/galaxy meets the interstellar space, there is a pileup of sorts. Pressure builds up and out light quanta now pops free of this system and sails off into interstellar space at the speed of C^2.

For those of you with a mathematical acuity, this means that light does not take thousands of light-years to reach us. Light from our nearest neighbor, Alpha Centauri, probably only takes about a day and a half to reach us. That whole light-year thing is just too much of a stretch for civilized man.

Now you wish an offer of proof; something tangible that you can either replicate or scribble in your little notepads. Okay.

Several years ago, I went camping and was lying there, staring up at the stars. I was in awe that light could travel so far without running into an obstacle or diminishing to nothing (just as a flashlight does after x amount of feet). Oh what could the answer be?

As I watched, I saw a manmade satellite as it orbited across the sky. It was then that I also spied a star slightly ahead of its position. It was obvious that the two were going to intersect and I began to wonder what was going to happen. Yes, I give you that, such a question was absurd. And the answer was more than obvious.

Then the most amazing thing happened. As the satellite approached the star, and it wasn't a particularly bright one, the satellite appeared to circumnavigate the star. I watched in fascination as the satellite appeared to alter course and arc around the twinkling star. Or was it twinkling? I really do not remember that part.

Within seconds, I knew the significance of what I had witnessed. And this is what transpired. Because that starlight had traveled so far, at such a high rate of speed, it altered the space ahead of it in such a way that space density was the phenomenon I observed.

When you think about this revelation, many things become crystal clear. If light that is traveling at C^2 collides with matter at the fringes of our solar system, at just the right angle, it would create gamma rays. So we now have an explanation for short and long term gamma ray bursts.

All of that energy from the starlight creates the seeming chaos we observe at the fringes, the Oort Cloud, and yes, even the debris we see strewn tangentially to the center of a galaxy.

See, we really do not need spooks and goblins. The activity seen around "black holes" and so-called Event Horizons, is really nothing more than light waves which have attained speeds greater than 186,231mps.

In the beginning of this book, I lambasted three people who received a Nobel Prize for Hubble's work; specifically, they alleged, like Hubble, that the further away an object is from us, the faster it is moving. This, they assert, gives credence to the idea that the universe is expanding and the further away a galaxy is, the faster it moves away from us.

That whole thing is preposterous nonsense. With nothing to impede its acceleration, light continues to do so in interstellar space. Therefore, it is the light that is moving faster and not the galaxies, themselves. Frauds, charlatans, phooey.

If you are going to award somebody a Nobel, don't do it because they are Jewish; give it to someone who truly deserves it. And stop ripping off famous men like Hubble. It makes you look weak and stupid.

Don't Have a COW, Man

Isn't it amazing how much you can learn in such a short period of time when you have the right instructor? All you really needed was for someone to step in and make you look at it in a different light. Ha-ha.

The good news, for you, is that we are only just getting started. Don't fret; it only gets better and better.

Einstein based his Relativity on a faulty yardstick. He assumed that the speed of light was constant. In fact, the speed of light is very much variable. It can be super slow or it can be super fast...and all points in between. Sorry Al.

Next on my big shot hit list, so to speak, is the always brilliant Isaac Newton. But Sir Isaac had one fatal flaw---he assumed that gravity is a force of attraction. In reality, there is no such thing as a force of attraction...unless one is talking in terms of intellect or emotional response.

When one observes something falling, it is easy to assume that the Earth is attracting it. When one observes two magnets, it is easy to believe that one is attracted to the other. But what if I could explain all of that in terms wherein no force of attraction is necessary? What if I could explain the motion of astro-bodies without the need for anything remotely resembling attraction?

Well, hold onto your hats because we just jumped onto a new ride at Seven Flags! And let's start with something easy...those pesky ol' magnets.

We all know that if we hold dissimilar poles towards one another, the magnets seem to attract each other, right? And we all know that we can swing one magnet around so that two similar poles are in close proximity and they repel one another. Now, why do you suppose that is?

Action at a distance? Nope. Not hardly. Come on, man, think! By playing with the magnets, we have, albeit inadvertently, established that there is an energy source present. And you know what that energy source is doing? It's flowing.

You can go ahead and look for alternate explanations all day long; won't do you any good. The fact remains, there is a flow of power going through, or around, the magnets. Facing one way, they are in alignment and they seem to attract. Spun around, the two sources of force cause those forces to "collide."

Deal with it.

So now we come to a much bigger problem. You want to know how I intend to prove something as drab as a planet orbiting the sun. "Yeah, what about that, wise guy?"

Conventional science claims that Earth is falling towards the sun because of some unseen force of attraction. What a load of crap! Let me give you a formula for calculating planetary orbits. And the beauty of this simple formula is that it also works very well for calculating planetary speeds and distances for objects thousands of light-years (sic) away!

"Don't have a cow, man," is Bart Simpson's moniker. It defines the cartoon character. More than that, it provides us with a very important measuring tool.

$C_{max}/O_p = W_s$. C_{max}, as we all know, is the maximum speed of light within a closed system. Divide that by O_p , which is the Orbital period of the planet (and is given in

28

Earth days). This simple formula gives us W_s, which is the Solar Wind speed at the planet's orbital path.

For the Earth, we divide 186,000mps by 365.3 days and we get a solar wind speed of 509.17 miles per second. That is precisely the magnitude that was measured by scientific instrumentation. And we can do the same for any planet anywhere in the observable universe.

If I remember correctly, I saw where a planet in another galaxy was orbiting its sun every forty days. 186,000 divided by 40 equals a wind speed of 4,650 miles per second. Besides being extremely fast, that tells us that the galaxy is relatively new.

I suppose I better clarify that statement. When I refer to a galaxy as being new, I mean that it recently collapsed in on itself and exploded into a new position and system. I think, pretty much. all galaxies are the same age, as far as their total years of existence. They just get "reborn."

So what does the COW formula tell us about planetary motion? It tells us that the sun is NOT attracting us. In reality, the solar wind from the sun is pushing us around sol. No gravitational attraction necessary.

Another thing that COW gives us is a glimpse into that oddity known as space. The formula, itself, implies that there is a change in space so that 186k also becomes a coefficient of distance.

Another way of putting that is the further out we go, the less resistance matter encounters. In effect, it takes less energy to move a planet the further away from the sun. On some level, that seems counter-intuitive. Nevertheless, the available facts bear it out.

If any of that boggles your mind, then consider the following. Not only is there less resistance, space is directional within the closed system. Energy follows the path of least resistance.

Okay, I have gotten your attention now. Still, there are two very important questions that you wish for me to answer. In the first question, you want to know why, if there is no such thing as gravitational attraction, moons always seem locked towards their hosts. That is to say, why does the moon always show us the same face instead of rotating on its axis?

The second question is that you wish to know how I am going to explain what pushes a planet towards the sun? If the sun is not attracting us, why don't we just sail away from the sun?

My explanation for the first question is definitely going to fly in the face of conventional wisdom (sic). Scientists feel that the Earth and Moon are attracted to each other. Not only does this keep the moon from rotating on its axis, it hinders the Earth's rotation. These guts believe that, if we ever lose our moon, the Earth will spin at a much faster rate. Boy, did they ever miss the mark!

Actually, if and when we do lose our moon, the Earth will stop spinning on its axis...or, at least slow way down. Some of you are now thinking that this proves that the two bodies were attracting each other. You use the theory that the moons motion was pulling the Earth around its axis. The major problem with that theory is that the Earth rotates once every twenty-four hours and the moon's orbit is nearly thirty days long. It just doesn't jibe.

\   Physics 101, any motion is the result of a net force. Given that there are several forces acting on everything at any given moment, one force must be of a greater magnitude than the rest if we are to move something.

Physics 102, since the Earth is rotating on its axis, there exists a net force that is stronger than the others. Likewise, since the moon is not (relatively speaking), the application of force must be nearly uniform.

I am fond of saying that I can answer any question so long as you are able to adequately explain the question. In this case, you completely forgot what the question was, or you wouldn't have asked it. Take a look.

What forces do you see at work in our solar system? And, just for the hell of it, let's forget gravity for a minute (because it really has no bearing, at all, on the problem).

We see the sun. It emits sunlight which, as we all know, is radiated energy. See?

Now follow that sunlight to Mercury. What do you see? Why, you see a planet that is mimicking a moon. It rotates very slowly on its axis and pretty much keeps the same face towards the sun. Do you know why? No, not because of gravitational attraction. I already told you that it doesn't play a part in this action.

Next, I want you to take a peek at Venus. See all of that sunlight? And what is Venus doing? Well, for the most part, it is doing exactly as Mercury is. Have you figured out the solution, yet?

Now we get to the third planet, our planet, Earth. See all of that sunlight? Isn't that lovely? But wait, why are we rotating? What's different?

Oh drat, there's that pesky moon! Why doesn't it go away and leave us alone? Therein lies the rub.

Just in case you missed that, the moon is a net force. Mercury and Venus do not have any moons and so they remain relatively calm. But we have a moon that is roughly one-sixth the size of our planet. That is a lot of net force. But how does it work?

See all of that radiated energy? See how it is dispersed evenly all over the surface of any sphere in its path? Because that light is spread evenly all over the surface, there is no net force. Hence a body will rotate very slowly, if at all, on its axis.

Planets with moons have net forces because those moons reflect light, or radiated energy, towards their hosts. Because the moon(s) are so much smaller than the host, the light is not reflected evenly on the host's surface. This creates the net force that causes a planet to rotate.

Why don't moons rotate when light is reflected from their planets? Because the planets are so much bigger than the moons, light is reflected evenly across the surface of the moon(s) and there is no appreciable net force.

And you thought you were going to have to complicate things with gravitational attraction!

For the answer to your second question, I would like for you to take a look at the following picture:

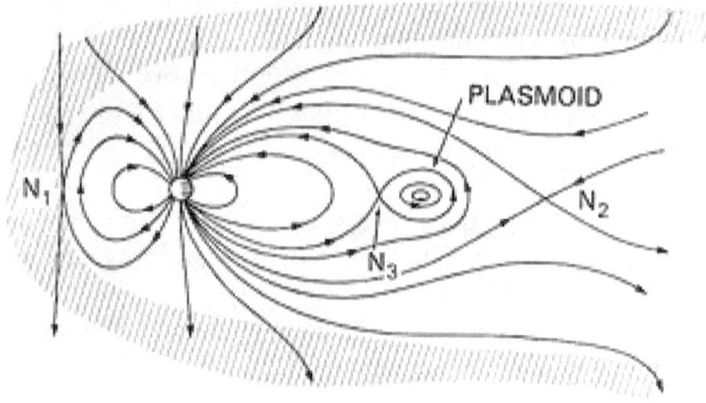

That teeny-tiny little ball towards the left of the picture is the planet (in this case, Earth). All of the squiggly lines are an artist's portrayal of the Earth's magnetic field. The answer to your question lies somewhere within this drawing. Can you figure it out?

Man you guys are hard to please. We draw you pictures, diagrams, we even eliminate the garbage that was cluttering your mind (by removing the dreaded gravitational attraction), and you still scratch your heads. Go ahead and take another look.

Have you figured it out now? Have you already forgotten everything we just discussed? Have you truly forgotten about net force and all of that?

Okay, you're right, I have been stalling. I wanted to give you the opportunity to decipher the clues for yourself. Call it class participation. Very well, let me get you off the hook. Take a look at the following picture:

Notice that I turned this picture sideways so as to better represent what I was trying to convey. If you compare this picture with the previous one, you will see the similarities.

A rocket moves because it has a nozzle that focuses all of that energy to the rear. Think of a sphere in space as doing the exact same thing. If you desire, you may attach a lengthy tirade about thermodynamics...certainly they play a part in the process.

When you look at the first picture I showed you, you see that the artist drew a wee bit of energy at the front of the Earth (nearest the sun) and an abundance of energy at the rear. You really do not need to be a rocket scientist to understand that when you have more energy at the rear than you do at the front, the object will move forward.

If that explanation is not enough for you, we could go into the topic of magnetic reconnection. Magnetic Reconnection is a physical process in a highly conductive plasma in which the magnetic topology is rearranged and magnetic energy is converted to kinetic energy (you know, the kind of energy that moves things), thermal energy, and particle acceleration. This description was quoted from the Wikipedia article on Magnetic Reconnection.

It is interesting to note the following peculiarity, also quoted from Wiki: "The qualitative description of the reconnection process is such that magnetic field lines from different magnetic domains (defined by the field line connectivity) are spliced to one another, changing their patterns of connectivity with respect to the sources. It is a violation of an approximate conservation law in plasma physics, and can concentrate mechanical or magnetic energy in both space and time."

I find the next paragraph in the Wiki article to be particularly helpful: In an electrically conductive plasma, magnetic field lines are grouped into 'domains' – bundles of field lines that connect from a particular place to another particular place, and that are topologically distinct from other field lines nearby. This topology is approximately preserved even when the magnetic field itself is strongly distorted by the presence of variable currents or motion of magnetic sources, because effects that might otherwise change the magnetic topology instead induce Eddy currents in the plasma; the Eddy currents have the effect of canceling out the topological change."

What all of that razzle-dazzle means is the solar wind blasts against the Earth's magnetic field and distorts it. The field condenses in the area facing the sun while the magnetic field stretches out behind the planet and forms a funnel or nozzle which contains the plasma field. The net effect of all of this is that plasma and/or magnetic energy gets converted to kinetic energy and this kinetic energy is what propels the planet towards the sun. The strength of this propulsion is largely contingent on the time frame of the process. Let's take a closer look.

An explosion is only powerful because of the rapid speed in which all of that energy is released. If we released all of the energy within an atom bomb within a split second, it is very destructive. However, if we released that energy in uniform packets, one every second, for a period of 2000 years, we might not even notice the heat energy. See?

Now I'll bet you that, several chapters ago, you were willing to wager that I could not produce an explanation, a viable explanation, that would replace the gravitational attraction that you have spent years learning about. And now?

People forget that those space particles in the solar wind are electromagnetic in nature. When those rub up against the electromagnetic field of the Earth, many things happen. The end result is that there exists a bellows-type action at the rear of the planet that pulses it towards the sun.

Scientists have discovered that whenever there is a solar flare, or a significant increase in solar activity, that it takes approximately 97 seconds after it passes us before the Northern lights, or Aurora Borealis, acts up. That clearly establishes a causal connection between the solar wind and the electrical activity in the Earth's atmosphere.

Moreover, since that action occurs after the wind has passed, then the energy causing the night-lights has to come from the rear of the planet (side away from the sun). And that, my friends, is a net force.

Just things to think about. A person should question everything...especially his superiors. Sometimes we give them more credit than they are deserving of.

My favorite saying is: The truth shall stand alone. That means it does not require the attachment of the name of a so-called expert. Nor does it need certified by a so-called higher board of education. The truth is the truth.

Another of my favorite sayings is: But what the hell do I know?

Most Shocking Secret of All

Each year, scientists make feeble efforts to justify their research...and their reason for being. Such was the case when these bozos decided that Pluto could no longer be a planet. I, for one, was dumbfounded.

After proclaiming that Pluto was being kicked out of the planetary lineup, these bozos spent months trying to find a way to justify the act. They could have used Pluto's very small size as justification; but they did not. Instead, they came up with some lame excuse about planetary pathways. It was foolish and completely without merit.

Now don't get me wrong; I'm not saying that Pluto is a planet. In fact, they were justified in giving poor Pluto the boot. Unfortunately, they also needed to give Mercury, Venus, Earth, and Mars, the axe. You see, none of those are planets, either.

I can hear the rhetoric now. You all think I done dove off the deep end. There's no hope for me. Am I serious? Let's take a look-see, shall we?

My definition of a planet is: "a large body in space that gives birth." It is just that simple. Wait, before you call in the white coats and commit me to a rubber room, you should hear the whole spiel.

Do you know how they make those little round lead balls for use in shotgun shells? They drip molten lead into a vat of water. Surface tension, and the like, makes the balls roll around into perfect little spheres. What does that remind you of? Spheres in space? Exactly!

You see those four giant gaseous planets out there beyond the Asteroid Belt? Those are frozen balls of methane. Care to tell me what those remind you of? Hint: you can't be a mother without one.

Give up? Very well, then, what you are looking at are eggs. You know, hard white shell. Gimme a break, will ya? I'm getting to it. I just want you to make a little effort.

In the birthing process, the egg is but half of the equation. What's missing is the spermatozoa. You all remember that little fella: elongated with a protruding tail, racing towards the egg?

Those of you who remember the previous chapter, know that I am fond of analogies. You already know what the egg is. You can choose between Jupiter, Saturn, Uranus, and Neptune. However, for my purposes, let's select Jupiter. No, not because it is the biggest but, rather, because of recent (relatively speaking) events.

In the 1990s, we were all privy to a spectacular event. The world watched in awe as the Shoemaker-Levy comet slammed into Jupiter. See that comet? Elongated with a protruding tail and heading right for our egg. That's not too terribly hard to envision, is it?

The part that you can't see is the molten ball of iron that is now rolling around on the surface of Jupiter. Just like kids making snowmen, someday it will get bigger and bigger. Think of all of that additional mass as Jupiter whirls around on its axis.

For those who do not know, Jupiter is the largest planet in the solar system. It is about 300 times the size of Earth and its day is about nine hours long. That is very fast and, looking back on the preceding chapter, one can surmise that it rotates that fast because it has many moons. And you'd be correct.

The four largest planets, the so-called gaseous giants, each have about a hundred moons. Each, not altogether! And, yes, they all rotate very fast.

Eventually, when our little snowball attains sufficient mass, rolls into just the right position, and conditions are just right, it will get flung off of Jupiter's surface to join its brothers and sisters in orbit.

By the way, all available data supports my hypothesis. All of the smaller round bodies within our system have iron cores. Of course they do! All have approximately the same density. And all either have water, or used to have water, on their surfaces.

Scientists will argue that most of the moons circling the large planets do not have the telltale ice of which I speak. Indeed, they will argue that our own moon does not have a layer of ice. Where did it all go?

Apparently, you have forgotten that these bodies have been around a very long time. We see moons around Jupiter that are in the throes of volcanic activity as we speak. Their icy surfaces are being covered with a layer of volcanic ash.

In other instances, meteors and such are bombarding these moons. How many craters does our moon have? A million? Two million? That's a lot of dust! So when they go to the moon and drill down a little bit, they will find water and my theory will be substantiated.

But we do not have to drill anything to prove the theory. One need only look at the Earth and Venus. The Earth is the ideal, at least for us, and is a prime example of how everything fell into place just perfectly. The methane ice melted and we became encased in Nitrogen (our atmosphere is largely that particular gas).

Methane is composed of both Nitrogen and Hydrogen. While the Nitrogen went into our atmosphere, the Hydrogen bonded with Oxygen to form water.

I believe that Venus was once like our Earth. It even had human inhabitants. Unfortunately, they did not pay attention to the eco-system and they killed off so much plant life that there was no longer enough oxygen to sustain it. The Nitrogen began to bond with the Hydrogen. Instead of Oxygen, they got Ammonia and acid rain. Are you paying attention?

Is this the thing that I said was going to be the most shocking thing you ever heard? No, but it damned well should be. Think of the ramifications of destroying the rain forests and throwing up concrete super-structures!

I think you have gained sufficient knowledge that I can now reveal the most shocking thing. You see that sun over yonder? Do you know what that is?

If you guessed that the sun is a furnace, give yourself fifty points. Do you know what fuels that furnace?

If you guessed Hydrogen, you almost had it. Man, you were so close. I was sure that you would have figured it out without my intervention.

Why do you think those four giant planets have so many moons? Do you really think it is just there to impress you? Nature does not work that way.

All of those moons, yes all of them, are there for only one purpose. Those are fuel pellets for the furnace we call Sol. Put that in your pipe and smoke it! :)

Doomsday Prophecy

If you were paying attention in the preceding chapter, you knew that I was not going to pass up the chance to expound on the subject of Global Warming. And I shall...in the next chapter. To me, the things I reveal in that chapter are even more shocking than the fact that we are just fuel for the sun. Meanwhile:

I wrote the following before "the End of the World" that they claimed the Mayans predicted. It was, and remains, one of my predictions. And you should listen up. It is, arguably, a very important piece of information.

So many theories abound which address the so-called Mayan Doomsday Prophecy (December 21, 2012). One of the classics is that the Earth's magnetic poles are going to reverse and such an event might be cataclysmic. I even entertained this notion for awhile. But, like everything else I do, I improved on it.

Science verified that the Earth's poles do, in fact, reverse every 25,000 years or so. They are at a loss to explain why that is so. They simply acknowledge that strata in the Earth's crust clearly reveals this happens. And it is damned hard to argue with forensics.

To understand what is taking place, please allow me to digress to the Garden of Eden. Adam and Eve gave birth to multiple children. Those children gave multiple births to other children. So on and so forth.

Eventually, the world becomes a very populated place. If you have ever stood in a crowded room (such as an elevator), you know how quickly the room becomes overheated. That is because each of us has a core temperature of around 98.6 degrees. That is nearly three times the temperature needed to melt ice.

Scientists have taken samples of ice at the polar caps by inserting a long tube and pulling it out. This shows them what was in the atmosphere over the course of thousands of years. What they found was a huge carbon footprint at about the time that a flood took place and/or a pole reversal occurred. And that is not coincidence.

Science tells us that we must reduce our carbon footprint in order to stave off polar ice melting and the subsequent flooding of the Earth. It is a nice suggestion but, like the sinking of the Titanic, it is inevitable. Reducing carbon buys us time, but not much.

I cringe every time I hear about people taking fertility drugs and spitting out quads and Quints. And Octomom totally blows me away. Idiots! Man's presence is the culprit. The faster we multiply, the faster we get to that flood stage and the ensuing disaster. Kind of hard for some people to envision but the reality is that octomom's eight kids will result in the extermination of three million people.

People live in the here and now without contemplating their actions or the impact on the future. As sure as I am writing this, overpopulation is a huge problem. And not just for the Chinese or Koreans. Look at the rest of the world.

We are all paying for these children. We are all paying with both the quality of our lives and the future of our existence. The ice will melt. And the poles will reverse. But not in accordance with generalized science.

When I say that the poles will reverse, I am not talking about a mere magnetic event. I am talking about the entire planet rolling over so that Antarctica is where the north pole currently resides. If you take out a pen or pencil and you rotate it clockwise, and you slowly turn the implement upside down, you will see that the pencil/pen is now rotating counterclockwise. Because the Earth is cutting through a magnetic field created by the sun, the magnetic poles are now reversed.

All of that is true enough. And the disaster could certainly result in a mass extinction event; thus fulfilling the Mayan prophecy. The only problem with such a scenario is that it is too precise. The Mayans had no way to predict world wars and the number of deaths therein. And so they could not accurately forecast when the overpopulation would reach its peak. No, only two events that I can think of would be so precise.

My personal theory was that the Mayans knew that the Earth's core is constructed of radioactive material. Such material decays over time in a very predictable time frame. It is possible that they predicted that the core would decay and die and life goes out on December 21, 2012. It is a very logical conclusion.

More recently, scientists revealed that there is a huge asteroid lurking in the shadows. They predict that the asteroid is going to pass very close to Earth; being somewhere between Earth and the moon. This is set to occur in February of 2013. That is a discrepancy of about two months from the date predicted by the Mayans. And that is much too close for comfort.

It is entirely possible that the interpreters of the Mayan calendar were off by two months. Or we could elicit many other factors to explain the discrepancy. In the final analysis, we must take the threat seriously.

The asteroid is estimated at about 150 yards across. A behemoth, to be sure. But not a planet killer. So what is the worst-case scenario?

Scientists might postulate that the asteroid could strike in the middle of an ocean and flood the whole planet. Or it could strike in the middle of a desert and saturate the atmosphere with dust particles. Neither of those would be very desirable and would be catastrophic. However, that would not be the worst-case scenario.

Worst-case scenario, by my estimation, would be if the asteroid collided with our moon. Such a collision would not obliterate the moon. But it could push it out of its orbit and send it rushing towards the sun.

Now, if you believe in classical science, then the planet will rotate faster. On the other hand, if you accept my admonishment that the planet will cease to rotate, then you will readily see that event as a potential extinction event. And the question arises, how long would it be before the Earth received another moon? And could anybody survive long enough to repopulate the planet?

There are a lot of unanswered questions. But they are questions that need answered now and not after the event.

There is one more possibility. It is one which I shudder to think about because it truly would be an extinction event. Let us suppose that this megalith misses both the Earth and the moon. Instead, let us suppose that it goes into the sun. Or maybe it strikes Mercury and one of them, or both, plummet into the sun. Ideas?

One massive solar flare accompanied by a heat wave so intense that it incinerates all in its path. Such a wave could blow away Earth's atmosphere and turn this ball of dirt into a cinder block.

Point is, that asteroid has numerous targets which could have immensely negative consequences for the inhabitants of the blue marble. It does not appease me to hear astrophysicists proclaim that the odds are that the thing will miss us. That does not pacify my troubled mind. Not one iota.

Top priority on everyone's minds all over the globe should be calculating, and recalculating, the next asteroid's path. If there is any chance at all that it will strike anything in the inner solar system then that needs to be addressed.

Granted, we are limited in what we can do to avert calamity. But anything is better than nothing. And I'll tell you the most frightening aspect. Of all the things that we might be able to do, forcing the asteroid to collide with the Earth might be the only viable solution which would ensure our survival. Huh?

If the asteroid strikes the moon, we die. If the asteroid goes into the sun, or causes anything to go into the sun, we die. A collision with the Earth is not at all desirable, but may well be the only solution as it is possible to live through it.

Caution must be exercised as a collision with the Earth might not be a solution if it causes a change in our relative position to our moon...or it changes our orbital path to any significant degree. A strike in the ocean would be preferable to a strike on land as it would spread some of the impact energy around a larger area; thus reducing the net force.

If the asteroid hit us in the right place, at the right angle, it could slow down our rotation or stop it completely. That wouldn't be desirable, either. See the quandary?

It is entirely possible that whatever the asteroid does, it could take until December 21, 2013 to reach fruition. In any case, we would be wise to accept the Mayan prediction and give leniency to the exact date. As a person who does not believe in coincidences, I do not feel that the presence of an asteroid within two months of the predicted date of extinction should be dismissed lightly.

Like so many others, I am used to thinking in terms of spatial qualities and we tend to ignore the temporal qualities. A lot of movies take this theme and run around with ingenious, albeit bizarre, solutions such as sending a demolition team to the surface of the asteroid so as to blow it to smithereens. But let's forget the spatial aspects for awhile and look at the temporal aspects.

Blowing up asteroids does not seem like a viable option with present technology. However, there does appear to be an emerging technology that might permit us to either accelerate an asteroid or to decelerate it. If we can get the asteroid to show up at a time when we are not there, then we do not need to blow it up at all.

In a previous chapter, I discussed, briefly, magnetic reconnection. At Princeton, they have a plasma lab where they are making some pretty interesting discoveries. A lot of them are counterintuitive to conventional wisdom.

Nasa, in cooperation with many other agencies, has put craft into orbit to observe the Earth's magnetotail and the plasma field. Specifically, they wanted to know what was taking place when magnetic reconnection took place out there. What they found out is that magnetic reconnection triggers Auroral activity (I.E. Aurora Borealis affects). There is a 96 second lapse between the reconnection and the visual anomalies.

When an asteroid travels through space, it cuts through magnetic flux lines which creates both ionic and electrical energy. This energy gets pushed to the rear of the asteroid where it forms a plasma field. The magnetotail contains the field and that propels the object forward.

Scientists will have an easy time understanding that, but will dismiss it as hogwash because of what happens when an asteroid, or comet, goes around the sun and heads back out into the solar system. Now, as we can plainly see, the magnetotail is in front of the celestial body. Why isn't it pushing the asteroid/comet back towards the sun?

Well, for one thing, the polarity has reversed. For another, it is now traveling in the same direction as the magnetic flux lines. The effect is that the magnetotail is now acting like a hot knife passing through butter.

Slowing down an asteroid may be as simple as placing a shield or deflector up on one side of it so that the containment field is breeched. Speeding up an asteroid is going to take a bit more work in that energy has to be applied within the plasma field without interrupting the containment wall. Perhaps a laser near the end of the asteroid and, possibly, contingent on the right angle.

I simply lack the empirical data necessary for the resolution of this particular problem. So it is best left in the hands of the experts. But they need to start looking into it.

Princeton's MRX (magnetic reconnection experiments) are a great place to look for answers. Let's hope that somebody takes the problem seriously enough to begin on it sooner than later. Or, maybe, it is better that we all perish. Time will tell.

Global Warming

I discussed a small portion of Global Warming in the preceding chapters. What I wish to do now is give you as clear a picture of this as I possibly can. This information is far more shocking, to me, than the revelation that we are fuel for the sun. In fact, if this information doesn't scare the hell out of you right now, you're not even human (and that is entirely within the realm of possibility).

People have sat in blizzards all over the world and scoffed at the idea that Global Warming even exists. It really is hard to believe when you are freezing your butt off. But we are going to dispel any lingering doubts that you may have.

I have already explained to you what happens when the oxygen level drops too low. With a shortage of oxygen, our nitrogen-rich atmosphere will begin bonding with hydrogen to form ammonia which, in turn, creates acid rain. What I haven't told you is where this hydrogen is coming from.

If you guessed water, you are absolutely correct. The nitrogen will draw the hydrogen from the water and the leftover oxygen will get extinguished by heat, fire, or utilized through the eco-system (birds, mammals, etc.).

There has been a lot of talk about our carbon footprint. This is not much of a problem until it gets to the point where all of that carbon expirates more and more oxygen.

In truth, the carbon problem is secondary to the main problem.

Do you know where that oxygen comes from? It comes from plants and trees and all sorts of foliage. Every time we cut down a tree, or rape the rain forests, or mow down shrubs so we can throw up a concrete structure, we deplete our oxygen levels.

Do you know that most commercial airliners routinely fly at 40,000 feet? Do you know why that is? Aside from getting them away from smaller aircraft, fact is, there is very little air up there. With no air, there is less wind resistance and the aircraft can fly further, faster, and using less fuel.

Forty thousand feet. That is not even eight miles. If you go eight miles straight up, there is no oxygen for you to breath. And that is all over the world!

In reality, the breathable atmosphere is only about two miles thick. Two lousy, measly, little miles; that's it! After that altitude, you will need to start supplementing your oxygen from another source. Two miles. A person can walk that in a matter of minutes.

Virtually every scientist in the world is aware how thin a membrane our life-sustaining atmosphere is. And the best they can come up with is to complain about carbon? Believe me, it isn't our carbon footprint that's at fault here. Sure, that isn't a good thing, but we need to get people more serious about maintaining the oxygen level of our thin blue line.

Forget the damned Spotted Owl; we are the endangered species. If we keep multiplying while, simultaneously, chopping down out trees, we suffocate. People in many parts of the country are already complaining about how hard it is to breath. We make light of the smog in Los Angeles and blame it on "those idiots." Fact is, we are all idiots who are destined to share that fate.

California recently passed some very stringent anti-pollution laws that target tractor trailer rigs. Problem is, they dumped that right in the laps of people who cannot afford to have eighty thousand dollars' worth of equipment put on their vehicles. I, personally, know of several owner/operators, and other small business people, who were forced into bankruptcy. That's great for the larger companies, but a damned lousy thing to do to a working man.

The issue with smog is exactly the same issue with dieting. People think that a diet is the answer to losing weight. In reality, the fault does not lie on gluttony, the real culprit is the refusal of the person to engage in enough exercise to burn off the calories he or she consumes.

A person can be the biggest pig on the planet. He or she can ingest five-thousand calories if they desire...just so long as they do enough physical activity to burn off five-thousand and five calories. I guarantee you will lose weight.

Now I am not advocating against diets. On the contrary. Our bodies consume fuel, just like a car does. You wouldn't dump water into your gasoline, would you? So what you eat is just as important as how much you work it off.

I know a family that regularly eats garbage. By garbage, I mean that they eat lots of chips, drink lots of soda, eat lots of greasy old hamburgers and fatty-rich steaks. They tend to avoid things like fruits and vegetables. Years ago, I predicted that these people would be suffering from clogged arteries and each has since had quadruple bypass surgery.

Now here is the ultimate in stupidity, even knowing the dangers inherent in their ridiculous diets, they continue to eat the same as before. And one has already experienced another heart-attack!

The point is, we keep dwelling on the idea of telling people what they can't do and they keep right on doing it. I suggest that we rehash strategies. Instead of telling people what they cannot do, why don't we start telling them what they can do?

I love going over to people's houses where they have innumerous plants as co-residents. The air in those homes is more pure and contains higher levels of oxygen. In addition, the aesthetics are not too shabby, either. A win-win.

We, as responsible human beings, need to contribute to our own welfares, as well as the welfare of society. If you are going to have fifteen kids and drive gas-guzzlers all over the place, you are using up a tremendous amount of oxygen. You should have to go out and plant two trees for every kid you have and four trees for every car. Plus, place foliage all over your house so you all can breath easier.

This is not optional. It is not subject to debate. This is a fact of life that nobody wants to think about, let alone discuss. We all prefer to think that we are individuals who can take away all of those resources without paying a price or replenishing them. We really need to grow up and start paying serious attention to these little details.

Global warming is a fact. Acid rain is a fact. People using up all of the oxygen in the air is a fact (especially in places like L.A.). How long are we going to bury our heads in the sand and hope the problem goes away or takes care of itself?

At this point, only an idiot would try to claim that the depletion of our breathable air is not a problem. Too many idiots are blaming carbon when, in fact, they need to blame our collective refusal to maintain adequate forests; forests which not only refurbish oxygen, but keep pollution levels down by purifying the air. So, yes L.A., you need more trees and less concrete. Nevermind the smog.

I believe that I have harped on the oxygen factor long enough. Let's take a look at the next problem on the list: water.

If we do not have respect for the air we breath, we have nothing but contempt for our water resources. While commercial airlines are plying the atmosphere with fossil fuels, we are busy dumping toxins into the oceans with thousands and thousands and thousands of commercial ships.

And how many millions of boats are running up and down our rivers, lakes, and what-have-you? How many of those boating enthusiasts ever stop to think that the bottled water they have also came from rivers and lakes and what-not?

Friction creates heat. Jet engines and boat motors create heat. Our bodies create heat. That is an awful lot of heat that we are pumping into our skies and waterways. Which brings us full circle to where I started...the polar ice caps.

Science has brought it to our attention that, in the next fifty years or so, the water levels along our coastlines are going to rise fifteen feet or more. Too bad, so sad, for beach dwellers. Right?

Too bad, so sad, for all of us. Let me tell you what these dipsticks won't tell you. As the ice melts and dumps the water into the oceans, it increases the amount of energy in the oceans. Anybody who has ever been sucked under by a swift tide, knows how strong those forces are.

We all know that the Earth is wobbling. Science blames that on the moon and on shifting magnetic forces. Nice try. Fact is, we wobble because of all that water (energy) in the oceans.

As we dump more and more water into the oceans, the energy increases and the Earth will wobble more erratically. Continental plates (tectonic plates, if you will) are floating on the Earth's mantle. All of this energy exuded by the ocean currents is going to wreck havoc with those plates.

Keep your eyes open for the increase in floods, the increase in earthquakes, and the increase in volcanoes.

These things are as inevitable as the stars in the sky appearing at night.

As if all of that isn't already bad enough, we have an even greater problem. All of that water that is getting dumped into the ocean is fresh water. So? So think about it.

In grade school and/or Junior High school, we learned that water is not a very good conductor of electricity. To make it a good conductor of electricity, we add salt. Are you following me?

As you know, oceans are primarily salt water. This means that they are great conductors of electricity. This means that the oceans are generating an electric current. Okay?

Next, we know that every electric current has an accompanying magnetic field. You still with me? This electric current induced magnetic field is what keeps our atmosphere from leeching off into space. And it protects us from dangerous cosmic rays; particularly those emanating from the sun.

As we dump more and more fresh water into the oceans, we dilute the salt and this weakens the electromagnetic fields of which so many creatures are dependent upon for their existence. Birds will have a hard time migrating, fish will get lost, and we'll be screwed.

In short, Global Warming is a very real, and very serious event. However, it is just a sign of a much bigger problem. We need to increase our forests and decrease our populations. And we need to do so today!

Biblical Prophecy

All of my adult life, I have heard numerous references to the number 666. In fact, I know way too many people who are actually superstitious as regards the number. Incredible.

Most of us know 666 as the number of the beast. This comes from the Biblical book of Revelations and is found in the back of the Bible as it is, literally, the last book of the New Testament. It is an interesting read.

I have actually read the Bible, in its entirety, some six times. I use to consider myself a bit of a Biblical scholar. Unfortunately, I have forgotten more than I remember. Not only is that a sad consequence of aging, it is also a consequence of having way too many other things occupying the space in my brain cells.

One thing that really stands out in my mind is the numerous references to a superior race of people. And I'm not talking about the Jews!

In the first book of the Old Testament, we read where the "sons of God saw that the daughters of men" were fair and took for themselves all that they desired. This is in keeping with many myths around the world where Gods descended from the heavens and demanded Earthly things. Indeed, we have all heard about the sacrificing of virgins to appease the Gods. Reckon, if I were a God, I would want virgins, too.

In the book of Ezekiel, he discusses, and describes, beings within an airship that has wheels within wheels. Could have been a flying saucer or, more likely, a helicopter or airplane.

There are numerous references to these beings and their crafts throughout the Bible. They make it a more interesting read; at least to me.

But the number 666 haunted me for the longest time as I was sure that it was some kind of a code. I was not alone in that department. Many researchers devoted inordinate amounts of time trying to decipher it. How funny that the solution was staring us right in the face---literally.

I'm always looking for the answers to mysteries and riddles. They keep my mind focused and keep me from getting stale. Of course it helps to have a very high I.Q. and that is not something that comes without effort. Think about it.

Fact of the matter is, hee-hee, we hear about 666 every single day of the week. Most of the time, we never give it a thought.

I apologize for laughing hysterically. I even discussed 666 in the preceding chapter...though I doubt that you noticed. If you are thinking that I am stalling, you're right. I wanted to give you the chance to kick it around a bit. I'll let you off the hook now.

The answer is carbon. The number 666 represents carbon. If you look it up on the Periodical Table, you will find that carbon is the number 6. Carbon has six protons, six neutrons, and six electrons. 6-6-6. Perfect!

The authorship of the Book of Revelations is attributed to St. Paul. There are many references to God telling Paul to give this message, and that message, to the people.

The Book of Revelations talks about apocalyptic events that are to befall mankind. Amongst other things, there are the four horsemen. These could represent the four forces known to the ancients; these being earth, wind, water and fire. Doesn't that make things more interesting?

So we pollute the air, we pollute the land, we pollute the sea, and we light fires everywhere. How fitting that these four forces manifest in four riders! Airplanes pollute the air, ships pollute the water, cars pollute the land, and all require fire to operate.

And then there is that pesky thing called carbon.

The Economy

The economy of the world cannot improve in any appreciable fashion until four things occur. First, real estate prices must come down to reasonable levels. It is lunacy for a mobile home park to charge more to rent a trailer space than what a person can rent a home for. That is blackmail based on the supposition that the homeowner does not wish to lose his investment. Blackmail. Nothing else. And the government needs to intercede on behalf of Mobile Home owners who are being held at bay by greedy landlords.

In the old days, one person worked to pay the bills and the other stayed at home to take care of the children, the housework, and all the other stuff. It is stupid to have an environment where both parents have to work just to pay their mortgage or rent payments. Instead of spreading the wealth, this merely puts the majority of the money into the hands of a few lucky enough, or corrupt enough, to own the property.

Simultaneous to the decrease in property values, we need to employ more people. Instead of handing out 600 billion to sustain Jewish and Italian Bankers, the government should have furnished 30 million people with salaries of $20,000. per year. Which was better for the economy; 30 Bankers dividing the loot or 30 million people dividing the loot? The answer is obvious.

The third thing that needs to happen is for the government to place caps on how wealthy a man can become and caps on yearly salaries. As a stockholder, I am shocked and appalled that the majority of the profits of a corporation goes to the man sitting in the top seat. That is ridiculous and is obviously embezzlement. In no legitimate business does the hired help make more than the person, or persons, doing the hiring.

Those CEOs are not doing anything that a grad student couldn't do for less money. In fact, most companies are really run by the people in the lower tiers. So why do the top dogs receive such lucrative rewards?

Fact is, CEOs are not reaping big bucks for running the company. CEOs make big bucks because they leak the company reports to their pals on Wall Street who, in turn, release the news to their pals in public office. The politicians look the other way because they are raking it in, too.

And the same thing applies to real estate. Everything the politicians are doing is to artificially inflate real estate because they, themselves own real estate. And that brings us to the fourth thing that needs to happen.

Number four on my list is to get rid of the corruption. Lawyers (and this includes all politicians) are buying up and/or building hotels, motels, and apartment buildings. They then convert these to subsidized housing wherein they bilk taxpayers out of double and triple the rents they would receive as a legitimate enterprise.

This subsidized housing scam goes hand in hand with the artificially inflated property values. In itself, the subsidized housing scheme inflates values and eliminates fair market values. Instead of having $300. a month apartments (in Salem, Oregon), we now have 800 and 900, if not higher. It is insane. And trailer space rent is the same as renting a house or apartment.

Motels, like the Cozy Inn, used to rent at 165 a week. Now they are something like 356 a week. And the motel is a dive compared to nice motels like Comfort Suites (which actually have rooms with saunas in them). When you rent a room at a dive and it costs more money than a room at a top notch motel, something is really rotten.

All construction contracts, these days, are corrupt if they involve the government paying for them. Prior to mandatory car insurance (which directly assaults the poor), car insurance was affordable. Now you pay through the nose. And if you refuse to pay the mafia for car insurance, they use the government to rip off poor people's cars/homes.

Hospitals struggled up until about eight years ago. Then they began to flourish. Why? Because the mob rushed in to buy them all up and fixed them so that you could get any medical treatment there. Why would they do that? Because they had already made a deal with their political allies to make health insurance mandatory. And don't think you aren't going to pay for that fiasco. They have already jacked up insurance rates (just like they did autos) so that you won't squawk when they start gouging uncle Sam (us taxpayers) for the insurance and medical costs. Up their's.

This kind of corruption needs to stop. Car insurance is something that should be tacked on at DMV because, in the end, the government is going to foot the bill for injuries anyway. Such insurance should be based on your earnings. It is a fact that poor people do less driving then their rich counterparts. They simply lack the money for gas. And it is historical fact that the people in the most accidents are the ones doing the most driving. It is inevitable. Why should somebody who is driving less than a hundred miles a year pay more than somebody who is driving a hundred miles per day?

Hospitals should not be privately owned. They should be owned by the government. As they are, they are bilking the taxpayer out of money for things (through medicare, etc.) that may not have even taken place. There have been recorded instances of men getting their tubes tied or even C-sections. And women reportedly getting treated for prostate cancer or vasectomies.

All of that money to build up every major hospital in the country came from taxpayers in some form. We should own those hospitals and eliminate making mobsters wealthy at our expense.

Times are changing and changes are needed. Mobsters raking in billions of dollars at our expense and then leaving us holding the bag for the damages has got to be a thing of the past. For instance, who foots the bill when citizens are thrown in jail for driving uninsured? And who foots the bill when citizen's cars are taken away for being uninsured? And who foots the bill when uninsured drivers are rearrested for driving on suspended licenses? Mandatory car insurance made the mafia rich but didn't reduce public debt any; it only increased it.

And all of this gambling. The taxpayers should own the casinos. Who foots the bill when gamblers lose everything they own? Who provides them with foodstamps? Who provides them with assisted living or subsidized housing when they lose it all? The mob rakes in all the profits and leaves the taxpayer the burden of cleaning up the mess.

And who pays for the jailing when a gambler loses everything and turns to crime to replenish his holdings? Or who pays for the damages of gamblers who steal things to support his habit? Who foots those bills? Certainly not the mafia.

Taxpayers in Salem, Oregon were bilked out of 30 million dollars to build a Conference Center which went to organized crime. An outfit calling themselves VIP, inc. received the conference Center and upscale motel free and clear. Must be nice.

Then there was the transit center/Marion Co offices fiasco. Millions of dollars poured into a construction scam that was doomed right from the get-go. It was structurally unsound and made of poor quality materials. Constant vibration from bus traffic weakened it and they condemned the entire property. It stands muted testimony to construction scams gone awry. And, of course, not one dime was taken away from the builder.

Awhile back, it was revealed that injured workers' claims were being routinely denied. Doctors were willfully and wantonly denying legitimate claims. Now that couldn't possibly have been because the hospital was raking in that money themselves? Like I said, I do not believe in coincidences. Funny how millions of dollars poured into constructing an all new hospital, replete with every service imaginable, and injured workers were receiving zip.

When the news leaked out, you know what our corrupt District Attorney did? He fined the insurance company (the taxpayers' insurance company and not Luciano's) almost a million bucks. So the solution was to fine the taxpayers and not the doctors. In fact, not one doctor was ever prosecuted or penalized. Isn't that marvelous?

The latest scheme to hit the airwaves is a 32.8 million dollar computer system upgrade for PERS. PERS is the retirement fund for State workers. It seems that some bright boy got the idea that these workers needed to know a more precise estimate of their retirement and they needed to know it today.

So somebody is stealing more than 30 Million dollars from the taxpayers on the flimsiest of excuses and neither the D.A. nor the Attorney General do anything about it? Makes you wonder how much those individuals are receiving from the con.

Think about it. There are hundreds of computer systems for way less than a million dollars that would handle every task in every arm of the government and simultaneously. A programmer could program it and keep it updated for under fifty thousand dollars per year. So where is all that money going? We know, don't we? Crooks.

When our forefathers devised the Constitution, they foresaw the need to have the government consist of three equal, but separate, powers. Each was to have veto power over the other two. However, that has all changed and the three have become a single entity who's only mission is to make each other wealthy at the taxpayer's expense.

So, what is the commonality? In a word, lawyers. Virtually everyone in politics today is an attorney. The diversity which our forefathers desired is now obsolete and perished. Lawyers have reinterpreted the Constitution to fulfill their greedy lusts. Case in point, eminent domain.

Eminent Domain was created as a means of obtaining goods and services in times of war. It basically said that the government could seize your assets in order to fend off an enemy. This was recently changed so that anybody can seize your property if they can show that such a seizure is in the best interests of the community.

In other words, if a mobster drove by your farmhouse and stipulated that the community would be better served with a shopping mall, then he can seize your farm. That is crazy. And typical of how the mob works these days. Let the government do the dirty work.

Several years ago, the mafia, er, boys on Wall Street, wanted to control OPEC. After all, the man who makes the most money is the man who either creates the news or knows what the news is before the rest of us. With no control over OPEC, the mob ran the risk of getting caught with their pants down. So they enlisted the aid of the United States military and political bigshots.

First, politicians convinced Venezuela to secede from OPEC. That infuriated the Arabs and they sent people down there to destroy oil pipe lines and to assassinate the top guys in Venezuela. When the U.S. was unable, or unwilling to help, Venezuela rejoined OPEC.

Then our corrupt leaders invented ways to place embargoes on OPEC nations in an effort to control them. That, of course, infuriated the Arabs and they sent people to retaliate. And the twin towers collapsed. Too bad they failed to take out Wall Street. But that is another tale.

Historically, the Jews have been in control of the media and of finances. Greenspan, Bernanke, and little Timothy Geithner come to mind. And who can forget their Italian cohorts? Men like Richard Grasso who ran around kissing cheeks and riding roughshod over a corrupt Stock Exchange?

If you disbelieve in the fact that Italian and Jewish mobsters are in control, read movie credits, watch CNBC or Bloomberg television. It isn't make believe. Well, the news they report is. Stay away from the Stock Exchange. It is the world's biggest casino and the chips they give you aren't worth the paper they are written on.

There is an old adage that desperate times call for desperate measures. Never before has that been as true as it is today. If three billionaires each coughed up 200 billion dollars, that would provide jobs for one third of all Americans (each getting $20,000 a year). That would be a damned good start. But then even that isn't going to work if we don't stop the stealing, the high rents, and all the other things that go with it.

Communism, by its very definition, is forcing people to do things they do not wish to do. Forcing people to buy car insurance so Jews and Italians can get rich is madness. Car insurance should be a choice. The economy was working well before car insurance. In no ways can an economy thrive when all of that money is going into the hands of a few.

The argument advanced for mandatory car insurance had been that the government was footing the bill for injured motorists who had been victims of uninsured accidents. Perhaps some of that was true. Problem is, the government is still footing the bill for injured people. Huh?

There is a mandatory $50,000 worth of coverage on a motor vehicle. On a particularly good day, thirty percent goes to the lawyers, thirty percent goes to the doctors and thirty percent goes to the victim. That's on a good day. In reality, the victim may never see a dime and will owe huge doctor bills. Only the lawyers make out.

Accident victims almost always draw some kind of public support (via foodstamps and welfare). Mandatory insurance makes that dependency longer, in most cases, because of the huge increase in people getting on the roster. This increase in recipients of public assistance is a direct, and indirect, result of people being penalized for not paying the mob for auto insurance.

People have to worry about shelter, food, clothes, and all of those things. Poor people are unable to pay the mafia for insurance because of survival needs. They lose their cars. This results in a lose of income from either not getting to work or not being able to drive associates to the store in exchange for a few dollars.

Uninsured motorists receive hefty fines which they cannot pay. They lose their homes. Now they are on welfare and/or drawing foodstamps. When they do not pay their fines, an arrest warrant goes out. Then they go to jail. Meanwhile, they lose everything they own.

They get out of jail and look for things to steal. They need money to get back on their feet. At some point, things get so bad that they assault someone in anger or to rob them; maybe worse. They know, as I know, that the government is corrupt and their troubles are because of that corruption. Being forced to pay the mafia owned and controlled insurance companies really pisses them off.

So they follow politicians home. Maybe they spray chemicals all over the place in the hopes that they cause cancer or birth defects; hopefully illness or death. Perhaps they get caught and go to prison. So another productive citizen, a hard-working soul who's only crime was taking care of his family instead of the mafia (and the mob's political allies), winds up behind bars. And when the man is released, he has little chance of getting a job because of his criminal record. Ditto for finding a decent home. And that man becomes another name on the list of people applying for low income housing (subsidized by taxpayers). And let's not forget the foodstamps.

And you want to expand this whole fiasco by handing the mafia a blank check for mandatory health insurance? Are you frigging crazy?

Here is the reality about the economy. The majority of the wealth is in the hands of the few. To get back on track, we need to get money into the hands of those who need it the most. And these are? Poor people.

Do away with mandatory insurance of any kind. Charge people a fee at the Department of Motor Vehicles when they go to license their car. This fee should be on a sliding scale where it is based on a person's income. Those making the most money should be paying a higher fee than those who have little income. This fee would be for things like doctor's bills, medicine, foodstamps, and helping people to keep their homes while they are injured.

The mob could still sell car insurance. This insurance could cover things like replacing the damaged vehicle, pain and suffering, and all those kinds of things. This should appeal to rich or middle class people.

It should be noted here that rich people do not, generally, purchase car insurance. Why? Because the law allows them to post a fifty thousand dollar bond instead.

Point is, we need to stop punishing people just because they are poor. We have not eliminated Debtor's prison; we are merely calling it something else. Fact is, people can, and do, go to prison just for being poor. And the taxpayers are footing the bill while the mob rakes in the gravy.

Stop the greed. Mandatory insurance is a communist plot hatched by corrupt mobsters and their political allies. Free America or else.

This chapter was written many years ago; long before I went on a five-year quest to discover the truth. Time to update it a bit.

I came out with a book called Broken Dreams and another called Twelve Stupid people. In the latter, I looked into the wrongful conviction of Frank Gable. What I discovered was the Jewish mob infiltrated my State and they were killing off people who were about to expose their criminal hyjinks.

The corruption involves the Governor, the Attorney General, the District Attorney, a slough of lawyers and a gaggle of Judges. In many instances during the so-called investigation, people's lives were threatened. Among these were an ex-cop an his wife. They had seen the murder victim alive at about a quarter past 7pm, when they observed him being chased by several men. Problem was, the prosecution was claiming that the murder victim got killed at 7pm when a lone assailant, Frank Gable, stabbed him at his car. Kinda hard for a dead man to be outrunning his assailants fifteen minutes after he's killed. Damned hard.

But these two people started receiving death threats via telephone calls. This is your first clue that something was wrong. You see, names of witnesses, etc, are kept confidential. The only way that anybody could have gotten the phone number, not to mention the names and testimony, was if it was a cop, or somebody the cop gave that information to.

In another incident, a State police officer pulled over Kevin Francke, brother of the murdered man, and threatened to kill Kevin if Kevin did not quit trying to ascertain who really killed his brother. Are you starting to get the picture that the police may have been involved in the crime? They weren't finished.

The next thing the police did was to select a man, any man, whom they thought they could convict. They selected Frank Gable because he was a lowlife but, more importantly, Gable was not a particularly bright fella.

Frank Gable had been a drug user and petty drug dealer. As the police gathered names of people Frank Gable associated with, they went around and told all of those people that Frank Gable was a rat (stool pigeon). One by one, after weeks and weeks of thinking about it, these people, all of whom were in prison, came forward to say that they either saw Frank do it, or Frank had told them that he had done it.

Now you are shocked by this behavior. Perhaps it would be less shocking if you knew that each of these people were felons who were sitting behind bars, and each thought that Frank Gable had been the person who ratted them off. In other words, each believed that Frank Gable was the reason they were locked up. And that was just exactly the police intention when they told all of them that story.

Here's the question you should be asking yourself: if it is so easy to prove that Michael Francke was not murdered at 7pm but was, in fact, murdered at least two hours later than that, how in the world could they succeed in convicting Frank Gable for the crime? That was the easy part.

The prosecutors saw to it that Frank received Robert Abel as his court-appointed attorney. Abel had long achieved two distinct features. First, he was the town drunk. So much so that he even went into court reeking if the stuff.

Second, Abel attained fame for being the man who got his clients convicted. That's right, instead of defending his clients, Abel willfully, wantonly, and deliberately got his own clients convicted; in this case, he purposely got Frank Gable convicted. How?

Robert Abel refused to let Frank testify in his own behalf. Right from day one, Abel told Frank Gable that Frank was not a credible witness. That is the same thing that the police had been telling Frank all along.

Oh, Mr. Abel did hire a professional who was supposed to make Frank a better witness. But the man was never allowed to meet with Frank. Abel had hired him for appearances only.

You know what else Abel did? He hired a private investigator to follow Kevin Francke (Michael Francke's brother) around. That's right. Robert Abel used taxpayer dollars to follow the murdered man's brother around. Can you guess why?

Kevin was trying to find out who really killed his brother. Just as important, he wanted to know the real reason why. The prosecution wanted to keep an eye on Kevin, but that would have looked too suspicious. Worse than that, if the State investigated and discovered something, and then failed to disclose that to the defendant, then there could have been a major lawsuit against the State. And the publicity would have been way more attention than any of the crooks in power wanted.

The fact that Robert Abel was trying to convict his own client did not go unnoticed by the men who were working for Mr. Abel. In fact, Robert Abel would get into a physical alteration with one of the defense investigators. And every single one of the men working under Abel contacted the court (Judge and prosecution) to complain about Abel's drinking and about Abel's intention to convict Frank Gable. Every one of them!

Frank Gable voiced his intent to testify in his own behalf. Frank wanted to tell the jurors that he was innocent. He made that known to the police who arrested him. He made that known to the trial Judge in numerous letters.

Bu the way, you know what the Judge did with those letters? Instead of reading them and helping Frank deal with a corrupt attorney, the Judge handed all of the letters back to Abel. Frank never did get to testify. Without hearing Frank's side of the story, the jurors believed that Frank was guilty. Of course they had to completely ignore the evidence. But that is how jurors' minds work. And Abel, and the Judge, and all of the prosecutors, knew that. It was all of their intentions that Frank be convicted.

I know you are chomping at the bit to know why. The murdered man had discovered the thefts of millions of dollars within his department. As head of Oregon's Department of Corrections (prisons), Michael Francke was slated to appear before the Oregon Legislators the following day. He was murdered and his files destroyed.

According to ex-Oregon Assistant Attorney General, Scott MacAlister, "they were supposed to make it look like a suicide." Indeed "they" were. Their plane had been to take Mike back to his office and use Mike's handgun to blow his brains out. Unfortunately for the bad guys, Mike managed to get away and ran to the highway in search of someone to help him. This is when Mr. Tucker (the ex-cop) and Mrs. Tucker almost ran over Mike and the two guys who were pursuing him. The time was 7:15pm.

At least two of Mike's coworkers were involved in the conspiracy to murder Mike. One of them contacted housekeeping and sent them home early (shortly before 7pm). Incidentally, that was something that had never happened before.

One of those coworkers told the police that he had seen Mike at about 6:45pm when Mike was going back to his office to call his estranged wife. Their plan was that this comment would make everyone believe that Mike called his wife, got rejected, and blew his own brains out. But Mike got away from them and, by the time they caught him, there were too many people running around the building.

Sometime in the night, somebody shredded nearly two-dozen bags full of documents. That took a lot of time, time that they wouldn't have had if housekeeping was still working.

Another reason that they did not want housekeeping around is because housekeeping would have told the police that none of those bags were there when they went home. What the bad guys overlooked was the fact that all of Mike's coworkers had been in meeting with him all day in that very office, and guess what? Yep, not a single bag was there.

In one of the oddest ironies of this whole crime, the man most responsible for getting Frank Gable locked away may very well have been the man who actually committed the crime. There can be no doubt that, at the very least, the lead witness knows who really did the killing.

Cappie "Shorty Harden's car was seen at the scene at precisely the time that Michael Francke disappeared. Coincidence?

When asked what he was doing on State Hospital grounds at 7pm at night, Harden replied that he was there to pick up his girlfriend. He claimed that she called him from a convenience store and told him to meet her at the Dome Building. However, when the cops went to ask his girlfriend about it, she denied having done that.

So the police contacted Shorty again. He managed to convince them to let him visit his girlfriend to explain to her that it was okay to admit that he went there to get her. Or, at least, that is the lie perpetrated by the cops. You see, both Shorty and his gal pal were in jail at the time.

Police procedure is that keep suspects separated. This procedure was implemented hundreds of years before so as to get at the truth. So the police deliberately violated police procedure in order to allow their favorite stoolie to get his story straight with his gal pal. Well guess what? Yep, you're right, after their consultation, which, by the way, could only take place by taking one out of jail to visit the other one who was in jail, too, the both of them got their stories to mesh. Sorta.

Shorty's girlfriend was under the age of eighteen. The parole office building (the place where Mike was murdered) was for adults. So, A) she had no reason to be there, B) she did not even know where that was or what it was called, and C) it was at least ten blocks from where she allegedly made the phone call.

The girl in question, was supposed to have called Shorty from a Pay phone at a convenience store that was at around 33rd and Center st. The place where they claimed she wanted Shorty to pick her up was at around 23rd and Center st. Shorty lived at around 7th and Center. This is important because the prosecution claimed that she waited around at 23rd and Center and then walked back to 33rd (instead of walking towards Shorty's house; which is her objective).

There is a phone booth at 18th and Center. It makes no sense that a teenaged girl would go to a dark secluded place in the middle of winter to wait for her boyfriend. Moreover, the parole office was on State Hospital grounds where crazy people live. You gonna believe that a girl who doesn't even know where that is, went there in the night, and waited for her boyfriend?

Not in this lifetime. And the story they concocted about her returning to the first store to call again is so full of holes...doesn't make sense. More importantly, it was proven that it couldn't have happened. Forensics, in this case, time, dictated that it could not have happened. And both witnesses have since recanted their incredible story. They admit to lying.

Question is, why did the jury even get to hear this story in the first place? At the point where Shorty's story did not match his girlfriend's, the police should have been looking at Shorty as the number one suspect. But they deliberately refused. Why?

Why let a murderer get off Scott-free and frame an innocent man? I asked myself that everyday. But then I have long known how corrupt the system is around these parts. My mistake was in believing that the organized crime was part of the Italian mafia when, in fact, it is the Jewish mafia.

I go into this in great detail in my book, Twelve Stupid People. It is incredible the lengths that they went to and the number of high-ranking officials that were involved. Indeed, the case reached all the way up to, and included, then Governor Neil Goldschmidt.

I am including a copy of a document that I call the Cory Memo. This is a five page document wherein Goldschmidt's legal advisor and confidante, Ms Cory Streissinger, writes the memo to inform Goldschmidt of his options in the Francke case.

The Cory Memo is a smoking gun, so to speak, in that it reads as a confession to multiple crimes. In it, amongst other things, is the admonishment that the Governor order an investigation and the declaration that the investigators not turn over any evidence that would link the murder to corruption in the prison system.

Cory also declares that the Governor must order the investigation or the Feds would come in and they did not want a federal agency looking into the crime. Those two things, in addition to the obvious Civil Rights violations, are Obstructions of Justice.

In the event that you are thinking that Goldschmidt is a fine upstanding citizen who would never stoop so low, let me give you a few tidbits. A few years ago, it was revealed that 30 year old, Neil Goldschmidt, while he was Mayor of Portland, Oregon, was having sex with his fourteen year old babysitter.

Some years later, Neil was busted, along with his wife and a friend, with conspiring to rip off stockholders in the whole Enron/PGE scandal. So this is NOT a man who refrains from felonious behavior; he embraces it.

For those of you who might choose to argue that such behaviors have no bearing on what he may have been doing as Governor, let's address that issue, too. While Governor, Neil absconded with nearly 200 million dollars of money that was supposed to go back to the taxpayers. You know what Neil did with the money? He embezzled it and gave it to the State Accident Insurance Fund, commonly called S.A.I.F.

Okay, I give. Now you want to know why would he do that and what would he gain? I'm glad you asked. You see, when Neil retired from his office, he went to work for S.A.I.F as a consultant. Got that? A frigging consultant.

Now let me ask you a couple of questions. Do you have any idea how much Mr. Goldschmidt was paid for being a Consultant for a company he illegally enriched by almost two-hundred million dollars?

In exchange for pilfering 176 Million dollars, Neil Goldschmidt was being paid forty thousand dollars per month! That is an outrage! Can anybody explain what the hell this man may have even known that was worth forty thou a month?

The State of Oregon pays experts in various fields between thirty thousand and eighty thousand dollars per year. That is per year. Not even Neil, himself, could provide a legitimate answer to this perplexing question. Nice to be a crook in office in this State.

But let's not leave Neil dangling on a noose by himself; let's look at some of the other criminals involved in the conspiracy to convict Frank Gable of a murder that was committed on behalf of the Jewish mob.

Cory Streissinger, yes, the very same woman who conspired with Governor Goldschmidt to coverup the connection between Mike's murder and the mob, went on to head Oregon Finances. Kinda of like putting the fox in the henhouse. Makes one wonder just exactly how much did she steal? And for whom?

You are wondering why I included her here if I cannot prove thefts occurred? And I shall answer you. Cory's deprivation of Frank Gable and Michael Francke's Civil Rights is a much bigger crime that any of the money she undoubtedly stole. Given her criminal history, one needs to look into her other acts. And we have probable cause to do so.

Additionally, in Racketeering, there is an item called Patterns of Racketeering. If we have a criminal acting up in this office, we can bet that she is acting up in her new one. And, just the fact that she was promoted to head Oregon Finances shows a pattern of rewarding bad criminal behaviors.

I have set the stage for other key players in the Conspiracy. District Attorney, Dale Penn, is another known mob associate who did, in fact, use his position to further mob interests. Boy did he ever! As a reward for his criminal acts, Penn was rewarded by, first, heading the Oregon Lottery (which, by the way, is a huge criminal enterprise in itself).

After screwing the taxpayers as District Attorney, Dale went on to screw us as head of the Oregon Lottery. We now have slot machines in every bar and people are allowed to cheat at bingo by using computers to beat out players without machines (this is illegal under O.R.S.167.167).

You know how they next rewarded Penn for his crimes? Damn, you're good. That's right, they made Dale Penn head of S.A.I.F. He is raking in millions of dollars per year (and you thought Goldschmidt was good!).

An interesting side note to this episode, while acting as District Attorney, it was discovered that several doctors were routinely dismissing injured workers legitimate workmen's comp claims (mine included). You know how many doctors Penn prosecuted? None, zero, zip. Instead of making sure that those doctors never ripped off another worker, Penn opted to take action against S.A.I.F. Yep, his solution was to protect taxpayers and working stiffs by having S.A.I.F. fined around three-quarters of a million dollars.

The mob wanted mandatory health insurance so they could rip off the taxpayers at will. They knew it was coming and so they, collectively, denied injured workers' claims and had that money rerouted to the local hospital. By the time so-called Obamacare came into affect, that puny hospital was on steroids. It has a department for everything imaginable. Thank you very much S.A.I.F.

If I keep going, this is going to be a book of unimaginable volume, so I am going to give you a couple more tidbits and call it food. As I said, I have expounded on the Frank Gable/Michael Francke case in another book. If you are really that interested, I suggest you read Twelve Stupid people.

Obviously the title, Twelve Stupid people, refers to the jurors in the matter of the State of Oregon Verses Frank Edward Gable. However, it should have read 23 Stupid people and one corrupt one. You see, there were actually two juries that convicted Frank.

In the Cory Memo, she referred to Dale Penn's expedited Grand Jury schedule. What she meant was that Penn would need an expedited schedule if they, the crooks, were going to need to hurry in the event that Penn was voted out of office. Remember, if they did not find a patsy soon, the Feds were going to step in and they could not allow that.

So the first jury to find Gable guilty was Dale Penn's Grand Jury. And I'll wager that you'll never guess how the crooks pulled that off.

Before you can figure out the how, you need to be aware that the Oregon Attorney General's Office is nothing but a den of thieves. And I do not mean that figuratively. They will rip you off at the drop of a hat. Unconscionably.

And so it was that the Attorney General had one of his boys, and Assistant Attorney General, go to the Marion County courthouse and sit on the Grand Jury that they would, ultimately, use to indict Frank Gable for a murder they all know he did not do. And I can, and have, proved that.

Thomas Denney, the mobster in question, was a Civil Servant. As such, he was exempt from serving on a jury. Moreover, as a trained prosecutor, he was biased and could not sit on an honest jury.

Now you are thinking that I could never prove that he was planted. The hell you say. Tell me, when a man is exempted from serving, and his residential data is confidential, how could he have even received a Summons to appear? In other words, the Court Clerk did not have privy to that information and, without that information, how could she send it to him?

I reckon you know what these outlaws did next? Yep, a corrupt Judge, another Jewish mobster, palmed a slip of paper and made a specific person Foreman of the Jury. Bet you can't guess who!

This is inexcusable. A trained prosecutor is a trained investigator. That is what the Grand Jury was supposed to do; investigate. So why didn't they?

Thomas Denney convinced the rest of the jury, by use of his official prestige and powers of persuasion, to indict Frank Gable for killing Mike at 7pm. No less than six of Mike's coworkers said that Mike was still alive at 9:30pm. Somebody please explain how a crook can convince 11 stupid people to ignore the facts of the case? Nobody, except a career criminal, excuse me, career prosecutor, could have pulled it off.

Convincing eleven dumbasses to indict was one thing, but how in the world did they intend to convince a full blown jury of twelve people to find Frank Gable guilty? As it turned out, that wasn't hard at all. All they needed to do was to confuse Frank by brainwashing him into thinking that he ought not to testify. "Aw, they'll never believe you Frank." they cried in unison.

Who were "they?" They were the police and the man's own attorneys. Just as the crooks got one of their own on the Grand Jury, they got one, two, of their own to represent the defendant. All they had to do was to keep telling Frank that he was not a credible witness and convince him not to testify. That's all.

The jurors never got a chance to bond with Frank. They desired to hear him proclaim his innocence. An innocent man would be yelling it from the rooftops...but not Frank. That was how the trial Judge, the D.A., the A.G., the police, and all the others, succeeded in framing a man they all knew, and know, is innocent. If there is any justice, they will all rot in hell.

The Cory Memo

009076

STATE OF OREGON   INTEROFFICE MEMO

GOVERNOR'S OFFICE

TO: Neil Goldschmidt   DATE: August 22, 1989

FROM: Cory Streisinger (CL)

SUBJECT: Francke Investigation

CONFIDENTIAL

This memo is a summary and follow-up to conversations you have had regarding the Francke murder investigation, any possible connection with wrongdoing at the Department of Corrections, and further actions that may be in order.

Where We Are Now

Investigations by the State Police and the district attorney's office have produced no evidence to support allegations that Mike Francke's death was in any way connected to events within the Department of Corrections. Indeed, the available evidence suggests the opposite: Nothing in Mike Francke's files or elsewhere indicate that he was working on anything other than his pressing budget problems, and Mike's family made no mention of the much-reported phone call about "organized criminal activity" when they had their initial interviews with the police.

In addition, we have no evidence of organized wrongdoing in the prison system today. Although not everybody connected to the 1986 investigation was ultimately convicted, the "higher-ups" associated with the investigation are no longer part of Oregon corrections. The State Police and the Department of Corrections are reasonably confident that, despite the uncomfortable level of drug use in the prisons, there is no corruption or organized crime at work.

Nevertheless, several factors may make an outside investigation advisable at this point:

-- The murder investigation has come to a virtual dead end. Dale Penn's current activities, which we've discussed, are unlikely to produce anything new even if an indictment ultimately results.

-- Regardless of the lack of evidence, it remains possible that Mike Francke's death was connected to wrongdoing in the Corrections Department. So long as the possibility cannot be ruled out, it may deserve investigation.

-- Due to their involvement in the 1986 investigation, the state's normal investigative arms -- the State Police, the DA's office, and Corrections -- are viewed as tainted by the Francke family and some members of the press and public. The same may hold true for the Department of Justice now that Scott MacAllister's name has been linked publicly with the investigation. Any investigation by these entities will be viewed with suspicion regardless of the facts. (███

81

- The Francke family's activities and the presence of Phil Stanford ensure that the issue will continue to receive attention by the press and public until some decisive step is taken to resolve it. Even identification of a potential murderer at this point is unlikely to stop the public's questions about possible links to wrongdoing at Corrections. We are starting to hear charges of "cover-up," and this may very likely be the response to any indictment.

- Legislators are starting to get pressured to do something about this issue. The leadership has said that they don't believe legislative action is appropriate or necessary, but they may not be able to hold this position for much longer if no executive-branch action is taken. An investigation initiated by the legislature would obviously be less desirable than one we might set up.

What Can Be Done

The suggestion you have discussed is the appointment of a special investigator or investigative team. The issues raised by this suggestion include the following:

1. What charge should the investigator be given?

There are three possible areas of inquiry: (a) investigation of the Francke murder, (b) investigation of possible wrongdoing at the Department of Corrections, and (c) investigation of any possible link between the two. Your initial thought was to limit the investigation to the third of these areas. You also suggested that the investigative team be directed to turn over to the appropriate authorities any information not related to a possible link.

In light of the recent focus on the 1986 prison investigation, however, I believe it may be wise to expand the scope somewhat. Obviously it would be inappropriate and an invasion of Dale Penn's jurisdiction to start a complete re-investigation of the Francke murder. And it would be premature to demand an investigation of wrongdoing at the Department of Corrections. But several newspapers have now alleged that they have "sources," with lots to tell, who are "afraid" to tell the State Police or the DA. If your investigative team is directed to turn its results over to the State Police or the DA in the absence of a Francke/corrections connection, these sources will still have an excuse to refuse to come forward — and we will be in the same spot we are in now.

I recommend that you direct the investigative team to do two things: (a) investigate any possible link between the Francke murder and the Department of Corrections, and (b) advise you as to whether further investigation of the Department is needed, and if so, by whom. If the team gives the Department a clean bill of health, so much the better. If the team recommends further investigation by the police or DA, you can order it. And if the team finds reason to recommend a further outside investigation, better you should know it now.

2. Who should lead the investigation?

You need someone of unquestioned integrity, who also has the investigative skills to take on a task like this. A former judge would be ideal. Herb Schwab's name met with general approval when it was suggested previously. ████████████████████████████ Jake Tanzer's name was also suggested, but this possibility is complicated by his involvement in the Portland Office Building lawsuit.

I have not been able to reach Stan or Ted yet to confirm other suggestions.

Author's note:

I did not black out the lines; that is how I received these.

3. Once investigator or several?

Your original discussion contemplated a single investigator, assisted by staff. However, you may also want to consider appointing a team of three with one of the three (such as Herb Schwab) in charge. This would allow you to include someone with law enforcement experience, such as Pierce Brooks, who Stan suggested but who would not be suitable for a lead position. I would also suggest including a community figure unconnected to state government, who cannot be accused of covering up. ▮

4. Who will provide the investigative staff?

This is a difficult problem. Obviously you cannot draw staff from the normal sources such as the State Police or the Department of Justice. The idea you discussed was to ask for the cooperation of local law enforcement, and to allow your investigator to select a team from law enforcement agencies statewide. This may well be the appropriate route to take. However, it requires that the investigative team be fairly familiar with law enforcement personnel statewide so as to make an adequate selection.

You may want to seek the advice of others on this point. For example, Mike Schrunk may well have some suggestions. Simply for the sake of appearances, I suggest that you not draw from Marion County (because of the ties to Dale Penn) or from the Multnomah County Sheriff's Department (because of the ties to Fred Pearce). It is also possible that you could get federal assistance, but I believe it is crucial not to turn this into a federal/FBI investigation if you want any assurance that it will stay within the bounds you have set.

5. How will the investigation be funded?

If you are to avoid charges of whitewash, the investigative team will have to be adequately funded. This may take an E-Board request, unless you believe the State Police budget can provide funds. If funds are taken from the State Police or other state agencies, the arrangement must be structured so as to preserve the investigation's complete independence. Investigative staff should not go on the agency payroll or otherwise be formally connected with the agency.

If you believe an E-Board request would be more appropriate, I would not forsee any problem getting legislative approval. The legislators are getting very nervous about this issue, and an E-Board vote would give them an opportunity to say that the legislature has responded. However, an agreement would have to be made with the leadership in advance not to engage in any extended discussion of details during the E-Board session. This should not turn into an opportunity for legislators to grill Fred Pearce about drug dealing in the prison. I would suggest that the E-Board action take the form of an appropriation to this office, not to State Police or Corrections, and that it be presented to the E-Board by Tom Imeson or myself without Fred, Emil or Dale being present.

I do not have any way of guessing what an appropriate budget would be. Stan Long may be able to advise on this; I have a call in to him but have not been able to reach him yet.

6. How public should the investigation be?

To some extent, this is a "damned if you do, damned if you don't" question. If you insist on the investigation being confidential, you will encounter charges of whitewash and cover-up. ▮
▮ However, if you ask that the investigation be semi-public (i.e., some access to records, frequent updates from the investigator), nothing will be accomplished and the various "sources" we keep hearing about will again have an excuse not to come forward.

I think the balance is an easy one; the investigation should be confidential. Without confidentiality, it will produce no results. I suggest that you direct the investigators to promise confidentiality to their sources, thus invoking the "confidential disclosures" exemption to the public records law. The investigators' product would also be subject to the "criminal investigatory material" exemption. You would need to make clear that you are taking this step to protect the confidentiality of sources who have felt reluctant to come forward thus far.

7. How should the investigation be set up?

I suggest that you issue an executive order. This would give the investigative team some legal standing, and would enable you to order other state agencies (such as Corrections and the State Police) to cooperate. It would also give you a mechanism to spell out the investigators' charge and functions in a formal way.

Alternatively, the investigative team could be set up as an adjunct to the grand jury. Although this idea may have some logistical problems, it is also attractive for a number of reasons. The investigators would have formal authority through the grand jury, rather than being outside the normal criminal justice process. They would be entitled to complete secrecy, they would have subpoena powers, and they would report directly to the grand jury when their work is complete. If you are interested in this possibility, it will obviously need more discussion with Dale Penn.

8. To whom should the investigators report?

Regardless of whether the investigators are officially connected to the grand jury, the grand jury should receive their report. The report should also go to you for any further action that may be necessary, i.e., any further investigations of the Corrections Department that may be called for.

9. What should your public position be?

As I see it, you will have to make some fairly fine distinctions in explaining why this investigation is being set up. You will obviously want to avoid any implication that the allegations about the Corrections Department are correct, but you also don't want to sound like you are pre-judging the outcome of the investigation. I suggest that you continue to state that you have seen no evidence of any connection between the Francke murder and Corrections, but that the allegations are obviously going to continue nevertheless, that if there is any truth to the allegations it's important that the truth be brought out, and that an outside investigation seems to be the best way of clearing the air one way or the other.

You will also need to be sensitive to questions about why you find it necessary to go outside normal investigative channels. If you opt for linking the investigators to the grand jury, this will be easy; the investigation will become part of the normal process. Otherwise, I suggest that you stress the value of a fresh look, and also say that you want to avoid any later challenges to the investigation on the basis of it being conducted by insiders. You will need to be particularly careful to characterize this investigation as an adjunct to the grand jury proceeding, not a replacement for it.

What Happens Next

Before taking any action on these proposals, I suggest that you talk to the following people to make sure they are comfortable with it:

(1) Dale Penn. You should discuss the relationship of this investigation to his grand jury proceeding, and you will also need to make clear that this is not intended as a criticism of his investigation thus far.

84

(2). Fred Pearce. You will need to talk to Fred about avoiding morale problems within Corrections as a result of this investigation. Corrections personnel need to understand that you have not joined the crowd of those accusing them of wrongdoing.

(3) Emil Brandaw. Apparently the State Police are considering bringing in a new team to take a fresh look at the Francke investigation. If you are going to call in outside investigators, this new team would be superfluous. You will need to discuss funding with Emil also, unless you would prefer the E-Board route.

(4) Dave Frohnmayer. He should get at least a courtesy call before any announcement is made, so that he won't be caught short in responding to questions.

Given Dale Penn's expedited grand jury schedule, all of this should take place fairly soon if you intend to proceed.

0854U
cc: Stan Long
　　 Ted Kulongoski
　　 Tom Imeson

Where we are today

My apologies for dragging you into the aforesaid case. I possess a genius I.Q. that allows me to see the truth in things that bog others down in an oceanic quagmire of misdirection. I could not, in all good conscience, reveal mysteries of the universe to you without exposing these criminals for what they are.

The Frank Gable case happened in Oregon. It began on the evening of January 17, 1989. Frank was convicted in early 1991. I, for one, am horrified that the government has gotten so corrupt that a man, a clearly, and undeniably, innocent man, can be framed for the murder of a man who was going to expose some of that government corruption. It is just unfathomable to me that we allow this to go on.

While this case did happen in Oregon, I discovered that it is happening in virtually every State in America. And it starts with the President and goes on down. The Jewish mob is everywhere.

The Civil Rights Act of 1964 prohibits both individuals and groups from discriminating against a person based on Race, Religion, etc., etc. 42US1983, 1985, and 1986, make it a felony to conspire to deprive a citizen of the equal rights and protections of the law. You know what that means?

It means that all of our elected officials today are felons. Every single one of them, and I mean from the President on down, belongs in prison.

What do you mean I can't prove that? That's the hell of it; it is the easiest thing in the world. But before I do that, let me ask you a rhetorical question. Would you vote in a man, or woman, who you knew, for an absolute fact, was an active, and involved, member of the Ku Klux Klan?

Obviously, if you are a Klansman, you would. But I am referring to the vast majority of Americans who are not affiliated with the KKK.

If you answered no, then I am going to call you a liar. You know why? Because that is exactly what you did, and do, every time you vote.

The Jews own and control the political parties and the candidates. You are free to vote on any person that they have selected for you. Every time you vote a person into office, you are voting in another Jew.

Many of you are of the opinion that I am racist. Not true. I am calling it as it is. There is a whole race of people who discriminate against the majority of Americans every damned day and we let them do it at will. What happened to the America that desired to be free? Where are all the patriots who declared that they were will to fight and die for this country?

Going to fight Jewish enemies in the Middle East, under orders from our Jewish Presidents and carried out by Jewish Generals such as Stormin' Norman Schwartzkopf, is not fighting for Americans; it is fighting for Jews.

So what do I have against Jews? The same exact thing you have against the KKK. There is no difference; no difference at all. The KKK is all for whites and the Jews are all for Jews.

Tell me, how can you sit in your chair and lambaste me for speaking out against discrimination when it is a fact, ;et me repeat that, a fact, that Jews refuse to marry outside of

88

their race. Can you keep a straight face while you tell me that refusing to marry outside your own race is not discrimination? Do you really think that Jew in office has your best interests in heart? Seriously?

We are the most ass-backwards country on the planet. Not only do we smile as the Jews take over America, and chuckle at their refusal to marry Americans, we gladly pick up weapons and go overseas to defend Israel. Something is really wrong with that whole scenario.

You dare call me racist when I am the one being discriminated against? Surely you jest?

It never ceases to amaze me how the Jews have taken over total control of the media and we buy into it. We hear, we see, we believe, only the things we hear on television or read in Jewish newspapers.

They own Hollywood. The studios are owned by Jews. Look that up on Wikipedia. The Oscars are the brainchild of the Jews and is awarded to Jews for acting, writing, directing, etc. In fact, the more nominations a movie receives, the more Jewish it is.

When Obama gave his State of the Union Address awhile back, he introduced a wounded soldier who had fought in the Middle East. Can you remember his name? Remsberg. Do you honestly believe that, out of the millions of soldiers available to him, he managed to pick a Jew? That was no accident. They control the media.

Years ago, I took a class in college called Transactional Analysis. In that class, we learned how to manipulate people. We practiced on individuals and on groups. It is a very interesting science. And that should scare the hell out of you. Imagine, a whole branch of science dedicated to screwing you.

We hear, time and again, how Germany screwed the Jews. What you never hear is the why. Do you really believe that the Germans hated the Jews for no reason? Do you really think that Jews have been booted out of every country in the world for no reason? Either all of those people, in all of those countries, were racist or the Jews, themselves, were. Pick one.

This is reality. Jewish control of the United States is also reality. Why do you think the Muslims hate us so much? They have taken over this country, as they have all countries before us, and shall not relinquish their stranglehold without armed conflict.

Did you know that, in many countries, a man faces the death sentence just for saying that he does not believe the holocaust happened. The only people who care that much about it are the Jews. The death sentence? Seriously? And America supports that? A man is put to death because of his opinion? That is ridiculously harsh and the United States, the country that is supposed to be free, backs that up 100%. Give that some thought.

Speaking of put to death, did you know that many Arab countries will kill their citizens if they join, or practice, Freemasonry. Why? Because they believe, as do I, that the Masons are in the control of the Jews.

My opinion is that anybody who seriously believes that their race, or even themselves, is better than me, is somebody who is predatory in nature. To support a Jew is equivalent to supporting the KKK. I abhor racism as it is nothing more than an extension of the whole temper-tantrum syndrome.

When you ask the average American who prints up the paper money that he has in his pocket, he will respond that the Federal government does. Most people truly believe that the United States Treasury is printing the money. But that is wrong.

Money is printed by the Federal Reserve Bank. That is a privately owned Bank. And can you guess who owns it? You are correct. The Federal Reserve Bank was formed, is owned by, and completely in the control of the Jewish mob.

Look up the Federal Reserve, Jekyl's Island, and World Bank, on the internet. Also research Rothschilds.

Ask the average American who we owe all those trillions of deficit dollars to, and he or she will respond with "the Chinese." Fact is, all of those trillions of dollars are owed to the Jews who own the Federal Reserve Bank. Anytime Uncle Sam wants more money, the Federal Reserve cranks it out and hands it to them.

Remember 2008? The Jews, galvanized by Ben Bernancke, informed Congress that, unless Congress approved trillions of dollars in bailout money, the economy was going to tank. And the idiots in Congress gave it to them. Do you know what those Jewish Bankers then did with our money?

The Federal Reserve took the money and divided it up amongst the twelve tribes of Israel, this consisting of the twelve branches of the Federal Reserve, which gave each of those branches a "surplus" of cash. Those branches then sent the "surplus" money back to the Central Bank and deposited it there. You see, there exists a law which allows those branches to do that. That law also states that they are to be paid x amount of interest on any monies deposited with the Fed.

Do you remember Congress raising hell because the Banks had kept the money instead of using it as they were supposed to? Now you know why. Wouldn't you like to print up a trillion dollars, loan it to the government, get it back, divide it amongst your family, and charge the government interest on the money---twice? What a sweet racket.

While we are on the subject of shady deals, let's talk about another law that is commonly used by these Jewish thugs to rip off virtually every country in the world. I'll bet that you did not know that there is a law that says that one country can loan another country billions of dollars "to stabilize their economy." Look up the IMF (International Monetary Fund).

This law is used everyday and not for the reason it was created. You see, for example, the United States loans, let's say, Great Britain, 100 billion dollars. At the time of the loan, the exchange rate is (for simplicity) one dollar for one British Pound. So a Jewish Bank in Great Britain gets 100 billion dollars which are then converted into 100 billion British Pounds.

After a month or so, the exchange rate becomes 1.25 Pounds to the dollar. The Bank in Britain then exchanges the 100 billion British Pounds for U.S. currency. They now have 125 billion U.S. dollars. They promptly send the 100 billion back, which pays off the debt they incurred. What?

That's right. These loans to foreign countries, loans that are supposed to stabilize their currency, are interest-free. All they are required to do is to repay the amount of the loan...with no adjustment based on either inflation or changes in the exchange rate. What a racket!

Enough. I have woken you up. Let's return to Physics as it is a much more relaxing pursuit.

I have done just about everything there is to be done in the cosmic world. I have purposely refrained from divulging too much about gravity. It isn't that I do not have anything to say about it but, rather, that I choose not to. Gravity is not the complex entity that science would have us believe. However, when you have one race wanting to dominate over all other races, it is something best kept out of their hands. Perhaps in a more civilized and socialized world.

My work has gravitated (I couldn't resist) towards the subatomic domain. I spent considerable time analyzing the Periodic Table. One thing that immediately springs to mind is the fact that it is not quite right. It seems that some silly human beings ruefully neglected to factor in such mundane things as acceleration, velocity, and rotational torque. Truth is, all things behave differently when they are moving. And nowhere is that truer than in a lazy man.

Yes, I'm funnin' with you. But just a little. By themselves, chemicals are lazy. Only when we apply energy to them, or remove energy from them, do we get superior results. A hydrogen bomb, and the sun, are two good examples. A chemical transformation is nothing more than a transfer of energy.

Chemists do not tend to think of chemical reactions as a transference of energy. In their minds, they are mixing chemicals together to create energy. However, as we all know, energy can neither be created nor destroyed.

So it is readily apparent that all of that energy must already exist within whatever chemicals we are conjoining. Carrying that logic a wee bit further, we only control matter by controlling energy. And that puts us on a par with gravity.

Wrong-thinking has hindered the advancement of science; particularly in the past 60 or so years. We constructed a Periodic Table to list all the elements of matter that we know to exist. Question is, have we listed all of the different elements of matter or have we listed the various ways in which energy interacts with matter? Extending that, is there really only one solitary unit of matter, or are there several different types?

My approach to this matter, no pun intended, was to state that each element possessed the same magnitude of energy. That is not the same thing as saying they possess the same quantity of energy.

Science lists, what they consider to be, the four known forces of the universe. We have electromagnetic, strong and weak nuclear forces, and gravity. Assuming that each atom has a little of each of these, let's use a hypothetical scenario.

We are all familiar with car batteries. Some of these are 40 amp, 60 amp, 80 amp, or whatever current rating they happen to be. However, the one constant is that each is twelve volts. So let us imagine that each of the elements has the same twelve volt charge. Further, let us assign each of the four forces an amperage rating. Just for argument, electromagnetic force is rated 4 amps, Strong nuclear force is 3 amps, weak is 2 amps, and gravity is 1 amp.

For the sake of simplicity, we shall assume that all four forces are 1 volt each. Now we can ascribe any combination of the four forces so long as they all add up to 12 volts. Again, for simplicity, let is say that hydrogen has 1 gravity, 1 weak nuclear force, 5 strong nuclear force, and 5 electromagnetic. What is that in amps?

Gravity equals 1 amp, the weak nuclear equals 2 amp, the strong nuclear equals $5 \times 3 = 15$ amps, and the electromagnetic equals $5 \times 4 = 20$ amps. So the total would be $1 + 2 + 15 + 20$, or 38 amps. At the other end of the of the Periodic Table, this array might be 5 gravity, 5 weak nuclear, 1 strong nuclear, and 1 electromagnetic.

This is vastly over-simplified for comprehension. In reality, some of these numbers are going to be much higher. In these cases, we would be using exponentials such as 10^6 or 10^{-10}, etc. It is much easier for the inquisitive mind of the unversed to omit such factors.

It is a tantalizing clue, is it not? I think it ironic that our primitive forefathers categorized things into four forces as well. Earth, wind, water, and fire. Myself, I believe the key lies elsewhere.

There exists a universal constant. It is the thing which has eluded mankind for eons. Einstein tried to make sense of it by using the speed of light as the universal constant. In my preceding dissertation, I utilized a constant charge (without denoting negative or positive). In this sense, I made the same blunder as my predecessors. I used a scalar property when a vector was really needed.

Yes, direction matters. It is all well and dandy that we have a given magnitude, but in what direction, and why?

For the answer, we must digress to the most simplistic thing in nature. I am, of course, referring to relationships.

We have three things to consider in nature. We have space, matter, and energy. If you wish, you may include a time coordinate. But, in reality, time is merely a measurement of the other three.

Of the three, space is the most important. Without space, there is no matter and no energy. Next on the list of importance is energy. Energy exists without matter, but matter cannot exist without energy. Therefore, low man on the totem pole is matter.

Fundamental laws exist which determine how each of the three components acts/reacts. Space may exist as a playground for the other two, but it acts differently towards each. Space constrains matter and exacerbates energy.

One could make the argument that all three behave as energy and so each is a succinctly different energy force. In that case, one might define space as the place where energy and matter balance out. However, if I elected to take that route I would say that each component is then a different perception of the same thing. And there is an inclination for science to take this very approach.

Truth be told, man will never, and can never, know what any of these things are. The best that we can hope for is to devise rules that help us to understand the role of these things in our universe.

Matter resists both space and energy. Energy exists to move matter, irregardless of space. And space exists as a regulator of both energy and matter.

Man has learned how to use energy and matter to manipulate space. We are still in our infancy and so we do not do that very well, as yet. We ooed and awed over jet propulsion but that wasn't very technological. We were, and are, merely using a series of explosions to propel something from hither to yon. How utterly barbaric.

When we lit of an atomic bomb, we ooed and awed before proclaiming that we were now masters if the universe. How infantile. The accomplishment ranked right up there with learning to make fire; we just made a bigger fire.

Birds migrate in very precise patterns because they tune in to the Earth's natural magnetic fields. They do not need satellites for GPS feats. Why can't man replicate those deeds?

For all his ingenuity, man is but a child in the universe. We believe that we are so smart, we award accolades for dumb theories, and we ignore those amongst us who have the answers. That isn't smart, or even wise.

Rich men proclaim that they are superior because of their wealth. Many of these men lack a third grade education and barely know how to tie their shoes. But they were born into wealth and their parent(s) taught them the rudimentaries of manipulating markets for capital gains. Superior?

We are entering a time where people are too busy struggling, as individuals, to even consider the planet as a whole. If I'm faced with starvation, how can I possibly be interested in minor annoyances such as Global Warming? Besides, if other people do not care what happens to me, why in the hell should I care what happens to them?

Well, I care. Despite overwhelming and magnanimous odds, I care. I care that the Earth is warming up because of overpopulation. I care that a race of people have taken over my country with the explicit intent to enslave us all. I care that people are starving to death while crooks on Wall Street are buying yachts. And I care that an innocent man is in prison for something that he did not do.

And I care about the Jews. They are like little children being misled by greedy tyrants who claim to be religious but totally lack in religious knowledge. No religion on Earth has ever tolerated the outright murder of people over petty stuff. And yet the Jewish warlords use it, religiously, as an excuse.

If you are truly enlightened, religious, or God's chosen people, then you ought to know that we are all equal. I have chosen my child to run to the store for a gallon of milk. I have chosen that child but not above all others. It is time for the Jewish people to come down to Earth and be civilized. You're not special.

The saddest thing about Jewish people is not their lack of ability to reason, the saddest part is their lack of wisdom. Jews have been persecuted for hundreds of years. They have been driven from every country on Earth (excluding the U.S.) and still they keep doing the bad behaviors that got them into trouble before. That's lunacy.

Yes, I feel sorry for the Jewish people, but not for what Hitler and hundreds of other Rulers have done to them. I feel sorry for them for what their own have done to them...and are still doing today.

You cannot move into somebody else's country and scheme to take it over for your people. America is about freedom for all people. You have no Right to come here and hire only Jews, marry only Jews, and rob, steal, cheat, murder, lie, and manipulate.

The average American is as much in the dark as is the average Jew. Everything is going to hell and they cannot figure it out. It isn't that the whole world hates Americans; the whole world hates Jews and we Americans have allowed our country to be taken over by Jews.

How would Jews feel if I moved into their houses and took over. What if I made them go to work, give me their paycheck, and I fed them bread and water? There's no difference. Well, there is. You see, the Jewish power lords have tried to own it all. They have stolen trillions of dollars from Americans to five to Jewish cohorts. Many Jews don't even know this, but they are going to pay when the day come.

Like it or not, America is exactly where Germany was immediately before World War Two. Exactly. The Jews own everything. The Jews only hire Jews. The average American is practically begging for table scraps. It isn't good.

Since the Jews have control of the military, whether it is through their Jewish President (Obama), or it is under all of the Jewish Generals they have in place, the Jewish power lords believe that another holocaust cannot happen. Care to make a wager on that?

I, for one, do not understand how reasonably intelligent Jews do not gather together and deal with the problem. Do you really want to die because some criminals took over the banks, the government, the media, and virtually every major business in the country? Do you really want to die so that a handful of Jews can rake in hundreds of billions of dollars?

No man should ever have more than a million dollars. It is an insane amount of money.

I can tell you what you need to do, and I can tell you why you need to do it, but I cannot do it for you. This is a train wreck that is going to happen. But it hasn't happened yet. Do you hear me?

As a sign of good faith, the first thing you should be doing is to get Frank Gable out of prison. No way in hell could anybody, let alone Frank, kill Michael Francke at 7pm when witnesses proved he was still alive at 9:30pm. You cannot argue with the facts.

Free Frank Gable!

Rogue's Gallery

Old age does a lot of things to a person's mind. Things that we used to know as true are no longer the case. Perhaps the worst of these is the idea that a person could not become President if they had not served in the military. Somewhere along the way, this law was abolished and that opened the door for foreigners to step in.

Personally, I am appalled that the top three people in line for the Presidency, this being Obama, Joseph Biden, and John Boehner, are all Jewish. Moreover, not one of them has ever served a day in the military...at least, not in the United States.

Obama didn't want to go. Joseph Biden is alleged to have had one asthma attack (In his entire life!) and that was his excuse for not going. John Boehner, what a worm, allegedly had a bad back and couldn't serve; pretty amazing when you consider that his back was well enough to play college football.

Let's take a look at the Jewish presence in America. Many of these people are people that Americans would swear are not Jewish. Take, for instance, Barrack Obama. Did you know that Barrack is a Hebrewlized word?

Barrack Obama sits on his throne and pounds his chest. Even when he surrounds himself with Jews, appoints nothing but Jews to office, and parades Jewish soldiers in front of us, people still doubt that he is Jewish. These people so want Obama to be Muslim (he isn't). Even when he announces that he is going to dedicate our military to defending Israel, even then, people still do not believe it.

Even seeing for yourself, you still have reservations. What's it take; a sledge hammer? Look:

Obama at Wailing Wall in Jerusalem

When you call him on his Judaism, you know what he'll tell you, he only went to Israel to be nice to the Jews. He wants the country to know that he is for all people. Oh really, I do not see him in any Muslim church...and there are many millions more of them then there are Jews.

I spent several years researching the migration of Jews into this country. One of the things that really intrigued me was the fact that anytime the Jews were persecuted somewhere, they came to this country. And the most amazing thing of all was that virtually every President was Jewish. That's right, all of them.

Let's look at some famous Jews, shall we?

George Bush, Jr.

Remember the scandal that arose when Bush junior invented the myth that Iraq had weapons of mass destruction so he could send troops to invade that country? Did you know he was really doing that to protect Israel?

9-11 happened because Muslims are/were tired of the United States's Jewish government. We gave nukes to Israel. Of course, considering the warring nature of the Jewish people, the surrounding countries are/were fearful. You would be, too. Look at Bush's father:

George Bush participates in a Hanukkah Celebration in the Old Executive Office Building on December 21, 1989.

George Bush Sr.

How about Jimmy Carter? Wasn't he Baptist? Glad you brought that up. Because of persecution and knowing that Americans would never sit still for an all-Jewish government, they joined other denominations. But they are still Jewish. Take a look:

Jimmy Carter

Most, if not all, of the Governors in the United States are also Jewish. Perhaps the most famous is this fella:

Arnold Schwartzenegger

What about Harry Truman? What about him?

David Ben-Gurion, Israeli Prime Minister, and Abba Eban, Israeli Ambassador to the United States, presented Harry S. Truman with a menorah in the White House on May 8th, 1951. August 5, 1951.

Harry Truman and cohorts

I want you to think about something for a minute. When terrorists attacked the Twin Towers, do you know why that was? Because that was a central place where Jewish mobsters congregated to manipulate money markets. And you already know George Bush junior's response. Instead of going after Osama Bin Laden, Bush sent the military in to thrash Iraq so Iraq could not go after Israel.

Franklin Delano Roosevelt did the same exact thing in 1941. When the Japs attacked Pearl Harbor, did Roosevelt send all of our might to crush Japan? Hell no. Roosevelt sent the majority of our men and supplies over to north Africa to defend, you guessed it, Israel.

Many regard FDR as one of the best Presidents ever. While Israel may boast that, I consider the man to be a traitorous dog. And I'll tell you why.

When the Japs invaded the Philippines, they were able to do so because FDR had refused to send needed equipment and supplies to the men who were defending her. Short of guns and ammunition, the GIs were forced to flee to a little island called Correigidor.

Some 75,000 men were trapped on that island. You know what Roosevelt did? He left them there to die so he could concentrate on saving his Jewish pals in Africa and Europe. Half of those men died when they were forced to surrender. The other half were tortured mercilessly by the Japanese soldiers.

Do not ever tell me what a hero Roosevelt was. They should have taken this piece of garbage out and shot him. When you care more about Jews who, by the way, hate the rest of us, and you leave the rest of us to die, you are a traitor. Period.

This is not Israel. If you cannot accept that all men are equal, then get the hell out.

Earlier, I mentioned the Michael Francke murder and how they had set it up to frame an innocent man. And Frank Gable is as innocent of the crime as you and I are.

The following document is Judge Yraguen's 2001 decision that denied one of Frank's many appeals. This Judge is, you guessed it, another Jewish mobster. This document, along with the Cory Memo, are all you need to solve the murder and to clear Frank Gable of the crime.

Pay close attention to the timelines and things that are out of the ordinary. As I said, they claimed that Frank killed Mike at 7pm, but numerous witnesses stated that the north portico window was not broken until some time after 9:30pm. The last thing Mike did was to put his fist through that glass and Mike had only minutes to live. No way in hell did he lie on that porch for three hours before breaking the window.

So what were some of these extraordinary oddities?

1. Somebody told housekeeping to go home early so that housekeeping would not stumble in on them as they held Mike hostage and, secondly, so that housekeeping would not be able to testify about garbage bags full of shredded documents that were not there when they left.

2. On the day before he was to testify before the Oregon Legislature, Mike is alleged to have gotten careless about his safety? They claim that Mike ran up to his killer (who was allegedly inside his car). They claim that Mike left his gun at home that day. And they claim that Mike forgot to lock his car after taking important (evidence) documents out to his car.

3. Mike had been going in and out of the north door all day long. It required a key to get in and out. That key was missing from Mike's key ring (which is why he had to bust the window).

4. Somebody turned off Mike's pager.

5. Somebody turned off the lights on the north porch so the bad guys could get in and out without being seen.

6. Cappie "Shorty" Harden, the witness most responsible for convicting Frank Gable, was seen at the murder scene at

the same time that Mike disappeared. To date, he has no viable explanation for why he was there.

7. Shorty claimed that he went there to pick up his girlfriend but, like Shorty, she had no reason to go there, and she did not even know the place existed.

8. The cops never even considered arresting Shorty. Instead, they violated numerous police procedures and arranged for Shorty to get out of jail to visit his girlfriend, who was also locked up, so he could get her to change her story (she originally told the police she did not anything about the lies he told the cops). That never happens in an honest system.

9. Police went around and made all of Frank Gable's associates believe that Frank Gable was a rat and that it was Frank who had gotten them all locked up for drug charges. This was a lie, fabricated by the police and succeeded in getting many jailbirds to come forward and tell lies that helped convict Frank. It is called Tampering with witnesses, obstruction of Justice, and violated numerous Civil Rights statutes.

10. One witness testified that Frank came to him with a garbage sack that had bloody clothes and possibly a knife in it. Allegedly, Frank gave him the bag to dispose of so that that evidence could not be used against him. Seriously?

11. The Jury was allowed to hear prejudicial and inflammatory testimony. Specifically, although the knife used in the murder is unknown (and most likely either a bayonet or pocketknife), police were allowed to lie and say that the evidence proves that Mike was killed with a kitchen knife from Frank's house.

12. Right from the get-go, Frank's chief lawyer tells Frank that he is a bad witness and he will not be testifying. An expert was hired to make Frank a credible witness but the expert is never allowed to meet with Frank. In other words, the witness was for show only.

13. It was so obvious to all of the defense team that defense attorney, Robert Able, was deliberately trying to frame Frank Gable that many of them contacted the trial Judge. Frank contacted the Judge also. All were chastised for their efforts. Or worse. Conveniently, at least one of them would later be murdered by the cops.

14. An ex-cop named Tucker, and his wife, started receiving death threats when they reported that they saw several men chasing Mike at around 7pm when he disappeared. How did the thugs get Tucker's name and phone number (unless the cops gave it to someone or called themselves)?

15. In the Cory memo, we read where Governor Goldschmidt wanted to hide any evidence that would tie the murder to corruption inside the government. Goldschmidt wanted to keep the Feds out. Goldschmidt had no concerns at all that Mike had been murdered. And Goldschmidt ordered the State Police to find a suspect fast. The order was signed August 22, 1989 and that is the same time that the police started framing Frank Gable. Coincidence?

16. The witnesses were all in prison or jail and each had been told that Frank Gable was a rat. If you pay close attention to what they say, you will see that Frank Gable would tell the cops something like "I was wearing sunglasses that night," and the next thing you know a witness pops up to claim that he saw Frank Gable in the vicinity of the murder and he was wearing, oh my, sunglasses. Remember, we are talking about two methheads who couldn't remember yesterday; let alone something that happened more than a year ago.

I have given you some things to look for as you read the Yraguen Decision (sic). There are many more. For instance, in one incident, the head janitor is outside and sees two men standing toe to toe. The taller of the two men then rushes to the main entrance of the Dome Building while the other guy takes off running. Yraguen claims that this is consistent with the confrontation between Frank Gable and Michael Francke. Yraguen purposely lies in order to justify his part in the conspiracy to keep Frank in prison. You want to know why?

Mike was the tallest man on the property. At 6 foot three inches tall, he was more than two inches taller than Frank Gable. "The taller man went inside the Dome Building." That is not consistent with the lie Yraugen perpetrates.

Additionally, the janitor had testified that the doors were locked. Only Mike had a key and could have been the only one of the two men who could have entered the building as it is factual knowledge that Frank Gable did not have a key to the front door.

The Jewish mob is alive and well in my State (Oregon). I guarantee you, it is alive and flourishing in your State. If this can happen to Frank Gable and Mike Francke, it can happen to you or I or your children, or anybody.

Kevin Francke's life was threatened by a State cop. Police killed one witness who came forward about Robert Able's corruption. I have received numerous threats; including one from Judge Greg West! Others came from the Oregon Attorney General's office. One even came from Homeland Security. Is this the world you want to live in?

I shall not stop until they release Frank Gable. The man, as you can see for yourself, is entirely innocent. His own lawyers convicted him. And, as if all of that wasn't bad enough, just to make sure that Frank Gable did not testify, they sent a mob lawyer, Karen Steele, in to talk him out of it. She even went so far as to marry him. After the conviction scheme succeeded, she divorced him.

Pass this book around. Send emails, go to chat rooms, write to your Congressman, Senator, and all of the rest. If enough of us step up, we can make these cretins give Frank

Gable his life back. Too bad that we cannot do the same for Mike. May he rest in peace.

Yraguen's Decision- pages 1 thru 21

IN THE CIRCUIT COURT OF THE STATE OF OREGON

FOR THE COUNTY OF MARION

****

FRANK EDWARD GABLE,	)	
Petitioner,		Case No. 95C12041
vs.		
		JUDGMENT ON PETITIONER'S REQUEST
STATE OF OREGON,		FOR POST-CONVICTION RELIEF
Defendant.	)	

****

INTRODUCTION, DELINEATION OF ISSUES

and

PRELIMINARY FINDINGS OF FACT

The above-entitled post-conviction relief matter came before the Court on May 1st, 2000, for trial to the Court on Petitioner's Request for Post-Conviction Relief under ORS Chapter 138. Trial continued on May 2nd, 2000, and was concluded on May 3rd, 2000.

The Petitioner Frank Edward Gable was represented by and through Ken Hadley, of Counsel. The Defendant State of Oregon was represented by and through Stephanie D. Andrus, Assistant Attorney General of the State of Oregon.

At the conclusion of trial, the Court set, at the Parties' request and in lieu of trial memorandums and closing arguments, a briefing schedule. The Parties were given until June 30th, 2000, to submit their

483

1   initial Closing Memorandums and until July 14th, 2000, to respond to the opposing Party's

2   Memorandum, if desired.

3   The Court also allowed Petitioner's request for DNA analysis of the fingernail scrapings and

4   blood on the clothing from the body of Michael Francke. Although the Court anticipated receipt of such

5   results before issuing Judgment in this post-conviction matter, the Court has been advised that it will

6   be several more weeks before the written report of DNA results will be released even though the DNA

7   testing was completed in November of 2000. This Court has determined that the filing of this opinion

8   with findings and conclusions and resulting Judgment not be further delayed, even though the results

9   of the DNA analysis have not been received. The Parties are given leave by this Court to move to reopen

10   these proceedings within thirty (30) days of the filing of such results if it appears that such DNA results

11   warrant any further action.

12   Petitioner Gable was convicted of six (6) Counts of Aggravated Murder and one (1) Count of

13   Murder. See Exhibit #5. Petitioner Gable was sentenced to life in prison without the possibility of

14   release or parole. The Victim in the case was Michael Francke, who at the time of his death was the

15   Superintendent of the Oregon Department of Corrections.

16   The allegations in Petitioner Gable's Third Amended Petition for Post-Conviction Relief

17   include the following particulars of inadequate assistance of counsel on which this case has been tried:

18   First Claim of Relief

19   That Petitioner was denied effective assistance of counsel under the Sixth and Fourteenth

20   Amendments to the United States Constitution and under Article I, Section 11, of the Oregon State

21   Constitution in that Defense Counsel Robert L. Abel and John F. Storkel failed to:

22   1. Meet with the Defendant and keep him advised of the State's investigation and the Defense's investigation.

23

    2. Meet with the Defendant before and during trial to plan an effective defense strategy.

24

    3. Give an Alibi Notice even though the Defendant had always denied being at the scene of the

25   death of Michael Francke.

26   4. Read the reports and discovery furnished by the State of Oregon.

27   5. Read the defense investigator reports, consult with them on an ongoing basis, and effectively

28   *Frank Edward Gable vs. State of Oregon Post-Conviction Judgment (a:\PCGable121500.JUD)*   Page -2-

1   use the information the investigators provided them with before and during trial.

2   6. Refrain, against advice of defense investigators, from turning over all information obtained
    by defense investigators to the State as discovery without reviewing to determine which portions
3   were discoverable and which should have been kept as non-discoverable work product or
    otherwise irrelevant, immaterial or not discoverable.

4

5   7. Effectively prepare for and effectively cross-examine the State's witnesses in general, and the
    following in particular:

6   a. Jodie Swearingen;
         b. Cappie Harden;
7   c. Janyne Gable;
         d. Mike Keerins;
8   e. John Crouse;
         f. John Kevin Walker; and
9   g. Kris Keerins.

10 8. Adequately investigate Timothy Natividad's and/or John Crouse's involvement in the murder
    of Michael Francke.

11

12 9. Adequately develop, investigate and produce at trial evidence that Timothy Natividad and/or
    John Crouse was the killer of Michael Francke.

13 10. Adequately develop, investigate and produce a qualified expert to conduct an examination
    of the automobile driven by Timothy Natividad on the night of the death of Michael Francke, for
14 blood or other trace evidence that would have connected him to the killing.

15 11. Subpoena and present testimony at the trial of the wife, family and others associated with
    Timothy Natividad that would have shown Mr. Natividad was the killer of Michael Francke and
16 not Frank Gable.

17 12. Adequately represent the Defendant by Robert L. Abel engaging in a pattern of excessive
    consumption of alcohol during preparation for the trial and during the trial.

18

19 13. Properly object and argue to the Trial Court that the Indictment referred to above was invalid
    because the Grand Jury Foreman, Thomas H. Denney, OSB No. 66034, was a career prosecutor
    employed by the Department of Justice of the State of Oregon and unauthorized persons were
20 present at the Grand Jury.

21 14. Object on the grounds of Ex Post Facto the Court's submitting to the jury in the penalty
    phase of the trial the possibility of the Defendant being sentenced to life without the possibility
22 of parole.

23 Second Claim of Relief

24   Petitioner Gable realleges Sections 1 through 8 of his First Claim and alleges that he was denied
    the right to testify on his own behalf under the Fifth and Fourteenth Amendments to the United States
25 Constitution and under Article I, Section 11, of the Oregon State Constitution in that Defense Counsel
    failed to allow him to testify in his own behalf in the guilt phases portion of his trial knowing that he was
26 unwilling to waive his right to testify in his own behalf.

27

28

Frank Edward Gable vs. State of Oregon Post-Conviction Judgment (a:\PCGable121500.JUD)   Page -3-

1

Third Claim of Relief

2 Petitioner Gable realleges Sections 1 through 8 of his First Claim and alleges that he was denied his right to due process under the Fourteenth Amendment to the United States Constitution and his right
3 to be heard by himself under Article I, Sections 10 and 11, of the Oregon Constitution in that the Presiding Judge, the Honorable C. Gregory West, failed to:

4

1. Give the Defendant an opportunity to be heard on his letter dated April 2nd, 1991, in which he
5 requested a hearing on his attorneys failure to consult with him and prepare for trial.

6 2. Give the Defendant an opportunity to be heard on his letter dated July 1st, 1991, in which he asked the court for relief based on several matters including lack of preparation, taking away
7 Defendant's right to testify, and odor on the breath of Defense Attorney Abel.

8 3. Postpone the trial or grant other appropriate relief when almost all of the defense investigators presented a letter to Robert L. Abel and John E. Storkel, that was delivered to the Court,
9 indicating that the Defense Attorneys were not prepared to proceed to trial.

10 ## Fourth Claim of Relief

11 Petitioner Gable realleges Sections 1 through 8 of his First Claim and alleges that the Court lacked jurisdiction of the Defendant because the Indictment on which he was charged was invalid for
12 the following reasons:

13 1. The Foreman of the Grand Jury, Thomas H. Denney, OSB No. 66034, was a career prosecutor employed by the Department of Justice of the State of Oregon, which was involved in the
14 investigation of the Francke homicide and should have been excused from service on the Grand Jury pursuant to ORS 10.050(2).

15

2. Oregon State Police Officer William Pierce was permitted to sit in on the Grand Jury
16 proceedings in violation of ORS 132.090.

17 ## Fifth Claim of Relief

18 Petitioner Gable realleges Sections 1 through 8 of his First Claim and alleges that he was denied effective assistance of appellate counsel under the Sixth and Fourteenth Amendments to the United
19 States Constitution and under Article I, Section 11, of the Oregon State Constitution in that appellate counsel failed to:

20

1. Properly and adequately argue all issues adequately raised by Trial Counsel. In particular, the
21 Trial Court's failure to allow Petitioner's attorneys to present evidence that Timothy Natividad and/or John Crouse were involved in the murder of Michael Francke.

22

2. Properly raise on appeal that the Indictment was in violation of ORS 10.050(2) because the
23 Foreman, Thomas H. Denney, OSB No. 66034, was a career prosecutor employed by the Department of Justice of the State of Oregon and because unauthorized persons were allowed
24 to sit in during Grand Jury testimony and deliberations in violation of ORS 132.090.

25 ## Sixth Claim of Relief

26 Petitioner Gable realleges Sections 1 through 8 of his First Claim and alleges that he was denied due process under the Fourteenth Amendment under the United States Constitution and his rights under
27 Article VI and his rights under Article I, Section 11, of the Oregon State Constitution for the following

28 *Frank Edward Gable vs. State of Oregon Post-Conviction Judgment (a:\PCGable121500.JUD)* Page -4-

1 | reasons:

2 | 1. That prosecutors Bostwick and Moore failed to disclose exculpatory evidence and failed to disclose plea agreements or promises to the following key witnesses:

3 |

4 | a. Jodie Swearingen;
    b. Cappie Harden;
    c. Janyne Gable;

5 | d. Mike Kerrins;
    e. John Crouse;

6 | f. John Kevin Walker; and
    g. Kris Keerins.

7 |

8 | Petitioner Gable requests that this Court reverse his conviction, vacate his sentence and release him from custody.

9 |

10 |

11 |

12 |

13 | ### SUMMARIES OF COURT PROCEEDINGS

14 | *Findings and Conclusions Regarding Evidence and Matters Presented by the Parties in State vs.*

15 | *Gable Trial including Guilt and Penalty Phases and Regarding Evidence and Matters*

16 | *Presented by the Parties in the Gable vs. State Post-Conviction Relief Proceedings*

17 |

18 |

19 | ### STATE vs. FRANK GABLE, MARION COUNTY CIRCUIT CASE NO.

20 | #### 90-C-20442

21 | 1. Summary of Pre-Trial Matters:

22 | a. A Marion County, Oregon, Secret Indictment charging Frank Edward Gable with six

23 | (6) Counts of Aggravated Murder and one (1) Count of Murder was signed on April 5th, 1990. See

24 | Exhibit #s 5 and 102. An Arrest Warrant was ordered on April 6th, 1990. Frank Gable was arrested on

25 | the Warrant on April 8th, 1990, at the Coos County Jail Facility in Coquille. See Exhibit #232, Pg. 7594.

26 | Mr. Gable was arraigned the next day on April 9th, 1990. Defense Counsel Robert Abel and John

27 | Storkel first appeared with Defendant Frank Gable on April 12th, 1990. See Exhibit #232, Pg. 18, and

28 | *Frank Edward Gable vs. State of Oregon Post-Conviction Judgment (a:\PCGable121500.JUD)*   Page -5-

1   Exhibit #106.

2   b. On July 10th, 1990, on Mr. Gable's pleas of "not guilty" a jury trial was scheduled on

3   the above-noted charges for January 7th, 1991. See Exhibit #232, Pg. 35. Numerous pre-trial motions

4   were filed by both Parties and resolved prior to the beginning of trial. Defendant Gable was also

5   scheduled to go to trial in Federal Court as an armed career criminal on December 12th, 1990. See

6   Exhibit #232, Pg. 146. Jury trial was continued in the State's case to March 4th, 1991. See Exhibit

7   #232, Pg. 156.

8   (1) A three-day Defense Team Meeting with Defense Counsel and the Defense

9   Investigators was held during January of 1991, during which the Investigators advised Defense Counsel

10   that they did not believe that the Defense could be ready by the March 4th, 1991, trial date. See Exhibit

11   #232, Pg. 235. This led to preparation of a letter to the presiding Circuit Judge Greg West which will

12   be dealt with in greater detail in the summary of proceedings. Defense Counsel filed a Motion and

13   supporting Affidavit for continuance and at hearing on February 4th, 1991, Defense Counsel advised the

14   Trial Judge that an additional thirty (30) days was needed. See Exhibit #232, Pg. 236. Judge West was

15   willing to provide additional investigative assistance, but he declined to move the trial date. See Exhibit

16   #232, Pg. 242.

17   c. Jury selection began on March 4th, 1991 [See Exhibit #232, Pg. 353 and thereafter] and

18   concluded on April 24th, 1991. Jury trial began with jury being sworn on April 26th, 1991. See Exhibit

19   #232, Pg. 5799. Opening statements were delivered on May 1st, 1991. See Exhibit #232, Pg. 5874 and

20   thereafter. The State's presentation of their case-in-chief concluded on May 31st, 1991. The Defense

21   began their case-in-chief on June 5th, 1991, and concluded on June 20th, 1991. See Exhibit #232, Pg.

22   9791. The State began and concluded its rebuttal evidence on June 21st, 1991. See Exhibit #232, Pg.

23   9821. No surrebuttal evidence was presented by the Defense. Closing arguments were concluded on

24   June 25th, 1991. See Exhibit #232, Vol. 72. The Court instructed the Jury on the law they were to apply

25   in the case on June 26th, 1991, and sent the Jury out to deliberate at 10:05am. See Exhibit #232, Vol.

26   73, and Pg. 10064 with regard to the beginning of deliberations. The next day, June 27th, 1991, the Jury

27   returned their verdicts at 4:35pm. See Exhibit #232, Pg. Frank Gable waived his right to be present

28   *Frank Edward Gable vs. State of Oregon Post-Conviction Judgment (a:\PCGable121500.JUD)*   Page -6-

1 when the verdicts were returned and was allowed to leave the courtroom. See Exhibit #232, Pg. 10070.

2 Frank Gable was convicted by Jury Verdict by a unanimous vote of all Counts. See Exhibit #232, Pgs.

3 10072, 10073.

4   d. The penalty phase of the trial was set to began on July 1st, 1991. See Exhibit #232, Pg.

5 10075. On July 1st, 1991, the Defense moved to continue the beginning of the penalty phase until July

6 2nd, 1991, which request was allowed. See Exhibit #232, Pgs. 10078, 10079. The State made its

7 opening statement, the Defense having reserved its argument, and the State began presenting evidence

8 on July 2nd, 1991. See Exhibit #232, Vols. 74, 75, 76. Mr. Storkel made the opening statement on

9 behalf of the Defense on July 10th, 1991, and began presenting evidence on behalf of Frank Gable. See

10 Exhibit #232, Vol. 76, and Pg. 10315 for beginning of opening statement. On July 11th, 1991, the Parties

11 presented their Closing Arguments, and the Jury was sent out to deliberate at 11:30am returning that

12 same day at 6:20pm. See Exhibit #232, Vol. 77.

13   e. The Jury answered the first three (3) questions presented in the affirmative and the

14 fourth question in the negative. See Exhibit #232, Pg. 10522. The Jury thus rejected the potential death

15 penalty. Frank Gable waived his right to a 48 hour delay in sentencing, and he was sentenced to life

16 imprisonment without the possibility of parole. See Exhibit #232, Pgs. 10525, 10526.

17

18 **II. STATE'S CASE IN CHIEF AND DIRECTLY RELATED MATTERS**

19   1. Factual summary of circumstances relating to disappearance of and locating the body of

20 Michael Francke:

21   a. Michael Francke, who at the time of his death was the Director of the Oregon

22 Department of Corrections, suffered a fatal knife wound during the evening hours of January 17th, 1989,

23 at or near the front circular parking area next to and West of the Dome Building, which houses the

24 administrative offices of the said Department of Corrections.

25   (1) There had been a light rain between 1:00am and 3:00am during the early

26 morning of January 17th, 1989, but no precipitation of any consequence after that time. See Exhibit

27 #232, Pgs. 6612, 6631-6633. The weather that evening was cold but dry. See Exhibit #232, Pg. 6267.

28 *Frank Edward Gable vs. State of Oregon Post-Conviction Judgment (a:\PCGable121500.JUD)*   Page -7-

1 At 4:00am on January 18th, 1989, the temperature was 48 degrees Fahrenheit . See Exhibit #232, Pg.

2 6268.

3   b. Michael Francke was last seen alive in the Dome Building walking back toward his

4 Office at approximately 6:45 p.m. on January 17th, 1989, by a co-worker, David Lee Caulley, who was

5 Administrator of Fiscal Services for the Department of Corrections. When David Caulley left the Dome

6 Building at about 6:50 p.m. he observed vehicles, all of which looked undisturbed, belonging to Michael

7 Francke, Elyse Clawson and Mary Blake still parked at the circle in front of the Dome Building. See

8 Exhibit #232, Pg. 7080.

9   (1) During the late afternoon of January 17th, 1989, Michael Francke had a

10 meeting with his Assistant Directors and members of his Executive Team Staff (Elyse Clawson, Dick

11 Peterson, Jan Curry, Evelyn Weeks, David Caulley, Fred Nichols, Tom Fuller, and Robey Eldridge). One

12 of the topics of discussion was the overview of the Department of Corrections which Michael Francke

13 was going to present to the Legislature the next day. The meeting lasted until approximately 5:30 to

14 5:45 p.m. Once during the meeting at about 4:00pm to 4:30pm, Michael Francke went out to his parked

15 vehicle and returned to the meeting. See Exhibit # 232, Pgs. 6935 - 6938. Whether the Francke vehicle

16 was or was not locked after this time was not established by the evidence.

17   (2) After that meeting, Elyse Clawson, Assistant Director for Community

18 Services, David Lee Caulley, Assistant Director of Business Services and Budget, and Jan Curry,

19 Assistant Director of Information Systems and Personnel, met with Michael Francke in his Office until

20 sometime between 6:25 to 6:45 p.m. (time estimates vary between members of the Executive Staff)

21 continuing the process of preparing Michael Francke for his appearance before the Legislature. See

22 Exhibit #232, Pg. 6938 - 6939; 7079, 8878. David Caulley had what appears to be the last conversation

23 with Michael Francke inside the Dome Building between 6:35 to 6:40 p.m. at the doorway to his office

24 and, upon the conclusion of that conversation, Mr. Francke told Mr. Caulley, as he was midway down

25 the hall in the direction of his office, that he was going back to his office to call his Wife. See Exhibit

26 #232, Pgs. 8878, 8879. As previously noted, Mr. Caulley left work from the front door of the Dome

27 Building at approximately 6:45 p.m. to 6:50 p.m. and did not see anything amiss with respect to Michael

28 *Frank Edward Gable vs. State of Oregon Post-Conviction Judgment (a:\PCGable121500.JUD)*   Page -8-

1 Francke's vehicle. See Exhibit #232, Pgs. 8882, 8884.

2   (3) After meeting with Michael Francke, Elyse Clawson then went to her Office

3 and conferred with Mary Blake, her Assistant, about her own appearance before the Legislature on some

4 pending bills until about 7:15 to 7:20 p.m. They went out the front steps of the Dome Building and

5 stood talking for a few moments. Elyse Clawson then noticed that the dome light was on in Michael

6 Francke's vehicle. See Exhibit #232, Pgs. 6939 - 6941, 8915 - 8917, 8935. She and Mary Blake

7 investigated and found the driver's door open and no keys in the ignition. See Exhibit #232, Pg. 6944.

8 Elyse Clawson closed and locked the door to the Francke vehicle. See Exhibit #232, Pg. 6944. Mary

9 Blake in the meantime went around the Dome Building in her car to see if there were lights on in in

10 Michael Francke's Office and then drove to the front of the Dome Building and went into the Dome

11 Building to Michael Francke's Office finding the door to the Office to be locked. See Exhibit #232, Pg.

12 6944 - 6945, 8920, 8938. They then tried to page Michael Francke's pager but received no response.

13 They then called Evelyn Weeks, David Caulley, and Robey Eldridge, and advised Security of what they

14 had found. See Exhibit #232, Pgs. 6945 - 6946, 8884, 8922, 8924. Elyse Clawson and Mary Blake left

15 the Dome Building again at about 7:45 p.m. See Exhibit #232, Pg. 6953.

16   (a) Security responded by sending Larry M. Hill to check the general area.

17 Mr. Hill drove around the Dome Building. Mr. Hill was unable to recall whether the porch lights were

18 on but did feel that there was light in the area although he was unable to see onto the North Portico porch

19 area. See Exhibit #232, Pgs. 9438 through 9441. Noe Pequeno, who is assigned to maintenance at the

20 Dome Building, believes that he turned on the lights on the North Portico when he locked the door to

21 the porch. See Exhibit #232, Pg. 9432.

22   (a) David Caulley was recalled by the Defense during the Defendant's

23 case in chief. See Exhibit #232, Pgs. 8877 through 8912. Additional details were given particularly

24 related to he and Dick Peterson's return to the Dome Building at approximately 8:30 p.m. to 8:45 p.m

25 in response to the Clawson/Blake calls. See Exhibit #232, Pg. 8889. Mr. Caulley talked by telephone

26 from the Dome Building with Robey Eldridge, Elyse Clawson, and Tom Fuller. Tom Fuller was

27 contacted at his pager number and thought that Michael Francke had a dinner engagement that evening,

28 *Frank Edward Gable vs. State of Oregon Post-Conviction Judgment (a:\PCGable121500.JUD)*   Page -9-

1 to wit, the evening of January 17th. See Exhibit #232, Pgs. 8897, 8899, 8906. David Caulley and Dick

2 Peterson left a note on Michael Francke's vehicle. See Exhibit #232, Pg. 8908. Dick Peterson was also

3 recalled by the Defendant during his case in chief. See Exhibit #232, beginning at Pg. 9257. Mr.

4 Peterson does recall seeing the door to the North Portico from inside Rm. #107 when he was about eight

5 to ten feet away from the door at approximately 8:40 to 8:50 p.m. as he was doing a general check of

6 the Dome Building. Mr. Peterson stated that the door appeared like a mirror and would lead him to

7 believe that the light on the porch wasn't illuminated. See Exhibit #232, Pgs. 9265, 9266, 9268. Room

8 #107 was only dimly lit from lights in the hallway of the Dome Building, and Mr. Peterson did not

9 observe any glass particles on the floor. See Exhibit #232, Pgs. 9276, 9277. .

10   (b) Mary Blake was also called by the Defendant during his case in chief

11 and also supplied additional details. See Exhibit #232, Pgs. 8914 through 8941. As previously noted,

12 when Mary Blake went around the Dome Building in her vehicle, she observed no lights in Mr.

13 Francke's office. See Exhibit #232, Pgs. 8920, 8938. She also recalls no lights on on the North Portico

14 porch. See Exhibit #232, Pg. 8922. Both she and Elyse Clawson were in the building at about 7:00 p.m.

15 but she related that it is difficult to hear any sounds from the North end of the Dome building from their

16 offices in the South portion. See Exhibit #232, Pg. 8932.

17   (c) Elyse Clawson was also recalled by the Defendant during his case in

18 chief and in essence simply reiterated her previous testimony. See Exhibit #232, Pgs. 9200 through

19 9211.   (4) The pager worn by Michael Francke was a Motorola pager which was

20 used in the vibration mode rather than the beeper mode. See Exhibit #232, Pgs. 6946, 6954, 6987, 6988.

21   c. Between 7:02 to 7:05 p.m. B. Wayne Hunsaker, a State Hospital maintenance man,

22 left the Dome Building via the basement tunnel door with steps leading to the front of the Dome

23 Building. See Exhibit #232, Pgs. 6841, 6843, 6849, 6871, 6878, 6880, 6983. Mr. Hunsaker was

24 heading toward the North 40 parking lot, and as he was walking North on the sidewalk toward said lot,

25 he heard "some sound of somebody being hurt***"; "kind of like somebody had their breath knocked

26 out"; "kind of a grunt sound." See Exhibit #232, Pg. 6881. Mr. Hunsaker turned back toward his left

27 and saw "two men facing each other." One of the men turned and went toward the Dome Building, and

28 *Frank Edward Gable vs. State of Oregon Post-Conviction Judgment (a:\PCGable121500.JUD)*   Page -10-

1 the other man turned and went in the opposite direction.

2   (1) The man hurrying East toward the Dome Building was about 5 feet 11 inches

3 to 6 feet tall, weighed about 175 pounds, had brown hair and was wearing a darker trench coat. See

4 Exhibit #232, Pgs. 6883, 6885, 6896. The man went up the steps of the Dome Building and disappeared

5 from sight.

6   (2) The other man was about 6 feet tall, weighed about 175 pounds, was between

7 the age of 20 to 40, had short black or dark brown hair, and wore a tan or beige trench coat about knee

8 length. See Exhibit #232, Pg. 6886. The man ran full speed to the street, hesitated for a moment,

9 crossed the street and then went behind a dumpster and generator across the street in the area of a

10 construction site in the hospital complex area. See Exhibit #232, Pg. 6887.

11   (3) The Defense during Defendant's case in chief also recalled Mr. Hunsaker,

12 whose testimony was consistent with his previous testimony. See Exhibit #232, Pgs. 9163 through

13 9199. Mr. Hunsaker did relate additional details including that as he was hanging up his keys in the

14 basement of the Dome Building the clock said 6:58 p.m. [Pg. 9191]; that he was about forty (40) feet

15 from the two individuals [Pg. 9175]; that for a fraction of a second the individuals were within inches

16 of each other's faces [Pg. 9176]; that he watched the incident for about a minute [Pg. 9171]; that the

17 incident could have occurred at about 7:02 p.m. or 7:03 p.m. [Pg. 9195]; that he saw no other cars or

18 persons coming or going in the area [Pgs. 9173; 9175]; that even after getting in his old beat up brown

19 Plymouth vehicle at about 7:05 p.m. or 7:06 p.m., he looked back and saw nothing [Pgs. 9174; 9181];

20 and that he wasn't even aware until the next day that the Dome Building had North and South porches

21 and stairs leading to those porches [Pg. 9183].

22   d. Between 7:05 p.m. to 7:07 p.m., Stanley S. Kudearoff and four other individuals, Pat

23 Boyd, Steve Laknas, Tyrone Williams, and a guy named Rocky, who were involved in an alcohol/drug

24 abuse treatment program, were on their way to a narcotics/alcohol anonymous meeting at the SOS Club

25 on Center Street. They passed in front of the Dome Building and noticed a parked white car with its

26 driver's door standing open. They looked into the car but noticed nothing unusual. See Exhibit #232,

27 Pgs. 6920 - 6923. Mr. Kudearoff recalls someone being over in the area of the North parking lot getting

28 *Frank Edward Gable vs. State of Oregon Post-Conviction Judgment (a:\PCGable121500.JUD)*   Page -11-

1 into his car when they passed in front of the Dome Building. See Exhibit #232, Pg. 6927, 6930. He

2 noted a cellular phone on the floor of the vehicle, and no flashing light on the dash of the vehicle. See

3 Exhibit #232, Pg. 6922, 6929. When the same group returned about two hours later retracing their

4 route, they noted that the vehicle was still parked in the same place but the door was closed. See Exhibit

5 #232, Pgs. 6924 - 6925.

6   (1) The State vehicle utilized by Michael Francke was a white Pontiac with

7 Oregon License # NNQ668. See Exhibit #232, Pg. 7909. During the ensuing investigation the cellular

8 phone located in the vehicle was checked by calling 911 and was found to be fully operational. See

9 Exhibit #232, Pg. 7911.

10   (2) Michael Francke had a marked parking space on the circle in front of the

11 Dome Building, and Mr. Francke's vehicle was observed by Evelyn Duffey in its marked parking space

12 at about 5:40pm on January 17th, 1989. See Exhibit #232, Pgs. 6682, 6683.

13   e. As previously noted, after receiving the telephone call from Elyse Clawson, Dave

14 Caulley called Richard Peterson, the Assistant Director of the Department of Corrections, and both of

15 them went to the Dome Building at approximately 8:20 to 8:25 p.m. Although they checked doors and

16 looked around the hallways, they didn't know how to activate the lights and thus were not able to see

17 into all of the working areas. They never checked the North Portico porch area of the Dome Building.

18 See Exhibit #232, Pgs. 7015, 7018, 7019, 7087, 7089. They left sticky notes on the door of Michael

19 Francke's Office in the Dome Building and on the driver's window of his vehicle and then left the area.

20 See Exhibit #232, Pg. 6949, 7019.

21   f. Shortly after midnight at approximately 12:42 a.m. on January 18th, 1989, a Security

22 Guard, Stephen J. Rubino, discovered a darkly-dressed body, which was subsequently identified as being

23 of remains of Michael Francke, lying in a awkward face up position at the base of the door on the North

24 Portico of the Dome Building. See Exhibit #232, Pg. 5975.

25   (1) The upper portion of the body was later discovered to be clothed in a shirt, suit

26 coat, pants, overcoat, scarf and cowboy boots. See Exhibit #232, Pg. 8259. The scarf was tucked under

27 one arm and was lying across the body. See Exhibit #232, Pg. 8399. The body was wearing a pager;

28 *Frank Edward Gable vs. State of Oregon Post-Conviction Judgment (a:\PCGable121500.JUD)*   Page -12-

1 there were some keys in a gutter at the body's feet; and there were glasses laying near the feet of the

2 body. See Exhibit #232, Pg. 8395.

3   (2) The body could not be observed from the sidewalk or from in front of the

4 stairs of the North Portico. Exhibit #232, Pg. 5956. The North Portico area was a covered porch area.

5 See Exhibit #232, Pg. 7871. Mr. Rubino observed a broken window pane and a hand smear on the door

6 near the door knob. See Exhibit #232, Pg. 5958. Mr. Rubino called Meril P. Craig, a Night Trade

7 Maintenance Worker, for assistance. Mr. Rubino and Mr. Craig approached to within about five feet

8 of the body, and then Mr. Rubino called his Communications Center and requested emergency medical

9 assistance and the police. See Exhibit #232, Pgs. 5959 and 5960.

10   (3) A hair found on the scarf was later examined and not identified although it

11 was similar to that of Bingta Francke, the estranged Wife of the victim Michael Francke. See Exhibit

12 #232, Pg. 8259.

13   (4) Subsequent examination of the damage to the clothing and blood on the

14 clothing and injuries to the body of Michael Francke revealed that the perpetrator probably used three

15 (3) knife thrusts: 1) one of which produced a puncture mark or slash of the overcoat which missed the

16 victim; 2) one of which penetrated and went through the victim's nylon overcoat sleeve, suit coat sleeve,

17 blue shirt, the victim's left arm, which arm was probably in a defensive position covering the chest, the

18 outer suit pocket which held business cards, the inner suit pocket containing a performance evaluation,

19 the shirt pocket containing note cards and then slightly into the left chest about an inch and a half; and

20 3) the fatal wound, described below, which penetrated the victim's shirt and the chest cavity a couple

21 of inches from where the left arm wound had penetrated the left chest cavity. The blade of the knife

22 used was pretty much oriented in the same direction, and it is likely that the same blade involved all

23 three (3) knife thrusts. See Exhibit #232, Pgs. 8355, 8356, 8358, 8359, 8367 through 8383. No other

24 trace evidence of any significance was located on any of the personal effects of Michael Francke. One

25 of the cowboy boots did reveal a scuff mark that occurred after blood had been deposited on it. See

26 Exhibit #232, Pgs. 8367-8368.

27

28 *Frank Edward Gable vs. State of Oregon Post-Conviction Judgment (a:\PCGable121500.JUD)*   Page -13-

2. Factual summary of investigation of site where the body of Michael Francke was located:

    a. At 12:46am, Salem Police Officers Dennis Fischer and Thomas Rousseau, who were in separate patrol vehicles, received a call to go to the Dome Building. See Exhibit #232, Pg. 6034. The Officers both arrived at approximately 12:49am, followed by Corporal Ed Kleinschmidt and Sergeant Gary Michel of the Salem Police Department. See Exhibit #232, Pg. 6075. The Officers were in turn followed by Salem Paramedic Tommy Newberry and Fire Department Personnel, William Groom, Willis Owen and John Bridges. See Exhibit #232, Pgs. 5960, 5961, 6035, 6037.

    (1) Several of those assembled went onto the porch and one of the emergency medical technicians, Tommy Newberry, checked the body for life signs but discovered no signs of life. See Exhibit #232, Pgs. 5961, 5962, 6043, 6103. Officers Fischer and Rousseau then taped off the immediate area including the vehicle parked on the circle in front of the Dome Building and awaited the Oregon State Police, who would control the investigation because the location was on State property. The taped off area subsequently grew larger and larger as the night progressed. See Exhibit #232, Pgs. 6040, 6062, 6063.

    (a) The first Oregon State Police Officer on the scene, Eric Karl Nelson, who arrived at approximately 1:50am, was able to identify the body as being that of Michael Francke. See Exhibit #232, Pg. 6154, 6155, 6850. Officer Nelson contacted his superiors and awaited the arrival of his superiors and crime lab personnel.

    (2) As noted above, William Groom and Willis Owen were among those individuals who first arrived at the scene of the murder. William Groom approached within two feet of the body of Michael Francke's body, and Willis Owen approached to within eight feet of the body. See Exhibit #232, Pgs. 6019, 6020, 6025. Both were wearing bunker boots.

    (3) Salem Police Officer Dennis Keena also arrived at the scene at 12:55am, checked the doors of the Dome Building and found nothing unlocked or amiss. See Exhibit #232, Pgs. 6126, 6135.

    (4) Richard Jensen, the Shift Supervisor of the State Hospital Communications Center, was also at the scene at approximately 1:15am on January 18th, 1989. See Exhibit #232, Pg.

Frank Edward Gable vs. State of Oregon Post-Conviction Judgment (a:\PCGable121500.JUD)   Page -14-

1   6589.

2   (5) The keys found at the feet of Michael Francke did not have a key to the door

3   Michael Francke was attempting to open. See Exhibit #232, Pg. 6856.

4   (6) The pager found on the body was in the "off" mode. See Exhibit #232, Pg.

5   6989.

6   b. Blood spatter and other physical evidence revealed that Michael Francke had been on

7   the sidewalk and had gone up the steps, attempted to open the locked door on the North Portico of the

8   Dome Building by breaking a pane of glass in the door, and died shortly thereafter on the porch without

9   gaining entrance to the Dome Building. See Exhibit #232, Pgs. 6516, 6145. The glass pane had been

10   broken inward and extended into the room about seven feet eight inches from the door. See Exhibit

11   #232, Pg. 6594. Skin recovered from the pane of glass revealed fibers similar to those of the victim's

12   overcoat. See Exhibit #232, Pgs. 8161-8161. There was no evidence of a struggle having taken place

13   on the porch. See Exhibit #232, Pg. 7888. The blood on the porch was Type O blood, which when

14   processed revealed that it was consistent with the blood of the victim Michael Francke. See Exhibit

15   #232, Pg. 7914.

16   (1) Oregon State Police Criminalists James Pex and N. Michael Hurley waited

17   until light to begin a detailed investigation of the scene. The Francke vehicle, a white Pontiac, Oregon

18   License # NNQ668, revealed no evidence of a struggle either on the outside or the inside of the car. See

19   Exhibit #232, Pgs. 6337, 6339. There was a sticky note on the driver's side glass left by Dick Peterson,

20   Assistant Director of the Department of Corrections. Exhibit #232, Pgs. 6254, 6290. There was a

21   footwear impression on the right rear bumper of the Francke vehicle, but subsequent investigation

22   showed that it belonged to David Caulley, a fellow Department of Corrections employee, who had on

23   a previous occasion assisted Michael Francke with his vehicle. See Exhibit #232, Pgs. 6603-6606. The

24   Francke vehicle was equipped with an operable Code Alarm security device including a shock sensor

25   but without a back-up battery. See Exhibit #232, Pgs. 6743 - 6745. A hair recovered from the driver's

26   arm rest was that of the victim Michael Francke. See Exhibit #232, Pg. 8258. The vehicle was later

27   examined for prints utilizing regular black powder, magnetic black powder, magnetic white powder,

28   *Frank Edward Gable vs. State of Oregon Post-Conviction Judgment (a:\PCGable121500.JUD)*   Page -15-

1 Rhodamine 6-G dye stain with laser examination and Luma-Lite examination with twenty-eight (28)

2 prints of value recovered eighteen (18) which were identified as being those of the victim Michael

3 Francke, his Wife Bingta June Francke, their Son Tre Francke, and Assistant Director Richard Peterson.

4 See Exhibit #232, Pgs. 8233.

5   (a) Lieutenant Pex conducted a demonstrative reenactment of the probable

6 sequence of knife thrusts which involved the initial cuts on Michael Francke's overcoat when he reached

7 across to get hold of his assailant who was apparently in his vehicle and slashed at Mr. Francke with

8 the knife; the assailant then exited the vehicle as the victim Michael Francke brought his left arm across

9 his body in a defensive position when the second blow was delivered causing the blade to go through

10 the left arm and slightly into the chest cavity of the victim; and then finally the last blow which was an

11 overhead thrust resulting in the fatal wound. See Exhibit #232, Pgs. 8403-8404. The second blow

12 could have produced the "ugh-type" of sound which would be expected when the lung was penetrated.

13 See Exhibit #232, Pg. 8404.

14   (2) The first sign of blood was located 95 feet from the car on a sidewalk leading

15 to the North Portico of the Dome Building. See Exhibit #232, Pgs. 6316, 6500. It was 34 feet from the

16 blood on the sidewalk to the top of the stairs leading to the North Portico of the Dome Building. See

17 Exhibit #232, Pg. 6501. A blood spot was also located fifteen (15) feet from the steps of the Portico.

18 See Exhibit #232, Pg. 7923. There were blood spots on the right side of the stairs. Exhibit #232, Pg.

19 6317. It was 28.5 feet from the top of the stairs to the body. Exhibit #232, Pg. 6501. There were blood

20 spots on the North Portico and a substantial amount of blood in the vicinity of the body. There were

21 blood droppings on the door and a bloody palm print on the door. See Exhibit #232, Pgs. 8395-8396.

22 The blood spatters reveal that Michael Francke, after receiving the fatal wound, moved along the

23 sidewalk in the direction of the stairs to the North Portico, went up and hesitated at the top of the stairs,

24 moved down the Portico, broke one of the window panes in the door leaving a small piece of his skin

25 on the glass, ended up on his knees before rolling to the position in which the body was discovered.

26 Exhibit #232, Pgs. 6521 through 6536. There was evidence of two cuts in the back sleeve of Michael

27 Francke's overcoat which were consistent with having been caused by rotating action of the overcoat

28 *Frank Edward Gable vs. State of Oregon Post-Conviction Judgment (a:\PCGable121500.JUD)*   Page -16-

498

1 | sleeve on the sharp glass after the pane was broken. See Exhibit #232, Pg. 8393. There was no evidence

2 | of any dragging of the body, of a struggle in the area, or of an assailant being in the vicinity where the

3 | body was located. Exhibit #232, Pgs. 6353, and 6521 through 6544.

4 | (a) Lieutenant Pex, in conducting a demonstrative reenactment consistent

5 | with the physical evidence found at the location where the body was found, determined that at one point

6 | Michael Francke was against the door and wall, slid down to his knees in his pooled blood and then

7 | rolled over onto his back in the position he was found. See Exhibit #232, Pgs. 8398 through 8401.

8 | (3) Dr. Larry V. Lewman, State Medical Examiner, established that the body of

9 | Michael Francke showed two stab wounds – one stab wound to the left ventricle of the heart, which was

10 | the fatal wound, and one stab wound which went through the left upper arm of and slightly penetrated

11 | the chest of Michael Francke as the left arm was held close to the chest. Both wounds were of the same

12 | general angle and direction. The fatal wound was left to right, front to back and above to below and was

13 | approximately 1" wide and 5 to 6" deep. Exhibit #232, Pgs. 6411 through 6417, 6454, 6460. The

14 | instrument was a thin blade approximately 1" wide and at least 5" long. Exhibit #232, Pgs. 6422, 6423.

15 | Death resulted from a progressive loss of sensorium (oxygen to the brain). Exhibit #232, Pgs. 6437,

16 | 6438. Michael Francke could have received the fatal wound close to his vehicle and then walked to the

17 | doorway of the North Portico of the Dome Building where his body was found. Exhibit #232, Pgs. 6437

18 | through 6441. The amount of blood found was consistent with death occurring as described because

19 | blood was draining into the left chest cavity and the amount of clothing worn by the deceased was also

20 | soaking up the blood. Exhibit #232, Pgs. 6441, 6451. Samples of blood taken from the remains of

21 | Michael Francke revealed that he had Type O blood. See Exhibit #232, Pg. 7914. There were injuries

22 | to the underside of Michael Francke's right hand [See Exhibit #232, Pgs. 8409-8410], and blood on his

23 | left hand and dirt on the left palm. See Exhibit #232, Pg. 8395.

24 | (a) There was no significant blunt force component of injury to the skull

25 | either through underlying hemorrhage or fracture (although there was a scrape on the side of the head

26 | apparently caused by a fall of the body during transport from the carrier). Exhibit #232, Pg. 6391.

27 | (b) Subsequent examination of the clothing, business cards, performance

28 | *Frank Edward Gable vs. State of Oregon Post-Conviction Judgment (a:\PCGable121500.JUD)*   Page -17-

1   evaluation, index cards coupled with the autopsy results and testing of all sorts of knives by Mr. Hurley

2   and Mr. Pex resulted in their opinion that the knife was: very sharp; single edge; flatback; contour of

3   tip not too pointed; five (5) to seven (7) inches long; one and one-eighth (1 1/8) or less in width; .04 to

4   .06 inch in thickness; with a relatively stiff blade having no abnormalities and blade which was not too

5   flexible. See Exhibit #232, Pgs. 8272, 8322.

6   (4) The door off the North Portico of the Dome Building was opened with a key,

7   but the door was not properly functioning on ~~December~~ January 17th, 1989. See Exhibit #232, Pgs. 6676, 6681.

8   The doorknob was designed to be operated with a key from the outside but with free egress from the

9   inside for fire exit purposes; however, the doorknob was rigid on both sides on December 17th, 1989,

10   necessitating the use of a key for both ingress and egress. See Exhibit #232, Pg. 6709 - 6712. In essence,

11   even though Michael Francke apparently broke the pane of glass and reached through in order to open

12   the door from the inside, the doorknob was rigid and could only be opened with a key.

13   (5) Limited evidence of shoe impressions, two partial impressions on the steps

14   and twelve partial impressions on the porch, were found, but all such footprint impressions, except for

15   one, the shoe impression on a blood droplet at the top of the stairs which was made by Stephen Rubino,

16   were attributed to impressions made before the murder occurred, to wit, probably when the porch was

17   being cleaned. See Exhibit #232, Pgs. 7831, 7833, 7858, 7864, 7866, 7874, 7875 through 7868. Other

18   shoe impressions located off the porch and steps were attributed to personnel associated with the

19   investigation, to wit: Gary Michel, Richard Jensen, Willis Owen, and William Groom. See Exhibit

20   #232, Pgs. 6609 - 6610, 6613 - 6615.

21   (a) The importance of the shoe impression made by Stephen Rubino

22   relates to time in that the impression was made at a time when the blood droplet was dry at least two (2)

23   hours after the droplet was deposited and was consistent with having been made from five (5) to five and

24   one-one half (5 ½) hours after the droplet was deposited. See Exhibit #232, Pgs. 7844, 7846 through

25   7849. No one associated with the investigation or the victim Michael Francke left any footprint

26   impressions on the steps or porch area of the North Portico. See Exhibit #7870. It should be recalled

27   that the Security Guard Stephen Rubino discovered the body at approximately 12:42 a.m. on January

28   *Frank Edward Gable vs. State of Oregon Post-Conviction Judgment (a:\PCGable121500.JUD)*   Page -18-

1 18th, 1989. See Exhibit #232, Pgs. 6949, 7019.

2   c. Michael Francke frequently used the side entrance to the Dome Building that opens

3 onto the North Portico and had used that door during the day on January 17th, 1989. See Exhibit #232,

4 Pg. 6679. Objects on the North porch are difficult to observe from ground level, and Michael Francke's

5 body was not observed by anyone until the Security Guard Stephen Rubino climbed the porch steps. See

6 Exhibit #232, Pg. 5956.

7

8   3. Factual summary of additional information regarding the investigation involving the death

9 of Michael Francke:

10   a. Oregon State Police Officer Loren T. Glover and David Calley of the Department of

11 Corrections did an initial check of the Michael Francke's house, located in Scotts Mills, Oregon, about

12 25 miles from the Dome Building in Salem, at about 4:50am on January 18th, 1989, and the house did

13 not show any signs of forced entry nor did anything appear to be disturbed. See Exhibit #232, Pgs. 6906

14 - 6909. Twenty-three (23) latent fingerprints were obtained at the Francke home with all twenty-three

15 (23) being identified without discovering anything of note. See Exhibit #232, Pg. 8226.

16   (1) Latent fingerprinting was also conducted in certain portions of the Dome

17 Building with twenty-nine (29) prints of value being obtained and eighteen (18) being identified, but

18 again nothing of note was discovered. See Exhibit #232, Pg. 8224.

19   b. The Michael Francke vehicle was examined for fingerprints and then processed for

20 blood. The outside window of the vehicle was weakly positive for blood using the orthotolidine test,

21 but there was not enough to test and James Pex surmised that it could just as easily been the remains of

22 a bug as of human blood. See Exhibit #232, Pgs. 7908 and 7909. Otherwise, no trace evidence or

23 anything else was found in the vehicle which related to the crime. The results of the fingerprint

24 examination was previously discussed elsewhere in this summary.

25   (1) One button which was missing from the suit worn by the victim Michael

26 Francke was located in the car. See Exhibit #232, Pg. 8386.

27   c. The Michael Francke residence located at 21058 Hazelnut Ridge Road near Scotts

28 *Frank Edward Gable vs. State of Oregon Post-Conviction Judgment (a:\PCGable121500.JUD)*   Page -19-

1   Mill was subsequently examined in detail but nothing of significance relating to the investigation was

2   found. See Exhibit #232, Pg. 7897,

3

4   4. Factual summary of matters directly relating to Frank Gable:

5   a. Personal circumstances involving Frank Gable before and after the murder of Michael

6   Francke:

7   (1) From December of 1985 through February of 1986, Frank Gable was a

8   Department of Corrections inmate assigned during daytime hours to deliver supplies through the Oregon

9   State Hospital complex area including the Dome Building. See Exhibit #232, Pgs. 7180, 7182. His

10   Supervisor was Nester Aujero. Frank Gable was among a group of inmates who were delivered to the

11   Dome Building in the morning and picked up from the Dome Building in the evening. See Exhibit #232,

12   Pgs. 7180, 7182. Frank Gable had access to and understood the layout of the tunnel system connecting

13   the various buildings and the layout of the various building involving the Oregon Hospital Complex

14   including the Dome Building. See Exhibit #232, Pg. 7185.

15   (a) Frank Gable participated in an inmate boxing program during 1984 and

16   1985 which was housed in the basement of the Dome Building. See Exhibit #232, Pg. 7215.

17   (b) Frank Gable was on formal parole on January 17th, 1989, under the

18   supervision of Duane Gamble of the Marion County Community Corrections Department. See Exhibit

19   #232, Pg. 7811. From various references throughout the transcript as well as from the statements of

20   Frank Gable himself, during late 1988 and early 1989, Frank Gable was heavily involved in the use and

21   sale of drugs, primarily methamphetamine.

22   (c) On April 9th, 1990, when Frank Gable was booked at the Marion

23   County Jail after being arrested for the murder of Michael Francke, Frank Gable was 6'1" in height,

24   weighed 175 pounds, and was 30 years of age having been born on August 28th, 1959. See Exhibit

25   #232, Pg. 7614. Frank Gable's appearance as of January, 1989, was noted on a video taken at that time

26   period. See Exhibit #232, Pg. 7699 [Exhibit #156 at trial] and was apparently consistent with the

27   description of the man observed running from the scene as described by Wayne Hunsaker.

28   *Frank Edward Gable vs. State of Oregon Post-Conviction Judgment (a:\PCGable121500.JUD)*   Page -20-

1   (2) Frank Gable and his Wife, Janyne Gable, resided at the Chandelle Park

2 Apartments at 3755 Hawthorne Avenue NW in Salem during January of 1989. See Exhibit #232, Pgs.

3 1988 and 1989. On January 18th, 1989, the Gables were given a 30-day eviction notice for excessive

4 noise and at a later time a 72-hour notice for nonpayment of rent. Other complaints regarding the Gables

5 had been received prior to the eviction notice. The Gables did finally vacate the premises on March

6 13th, 1989. See Exhibit #232, Pgs. 7222, 7223, 7689 and 7690. They then moved to the Skyview

7 Apartments in Salem; then the Clearview Apartments; and later moved in with Dan Walsh of Salem.

8 See Exhibit #232, Pgs. 7687 and 7689.

9   (a) Janyne Margaret Vierra Gable, the Wife of Frank Gable, worked at

10 Building 34 on the Oregon State Hospital Grounds. She had started working there in 1984 and worked

11 through the Spring of 1989 when she was terminated because of attendance problems. See Exhibit #232,

12 Pgs. 7677, 7683 and 7684. She met Frank Gable when he, as an inmate, drove a delivery route on the

13 Hospital grounds, and they were married in September of 1988. See Exhibit #232, Pgs. 7683, 7685, and

14 7703.

15   (b) Janyne Gable had keys to the State Hospital facilities, including the

16 Dome Building, of which Frank Gable was aware during her employment. The keys were never turned

17 in upon her termination of her employment. See Exhibit 232, Pgs. 7706 and 7708. Frank and Janyne

18 Gable were divorced in October of 1990. See Exhibit #232, Pg. 7683. At trial Jayne Gable identified

19 the keys she was issued, and Virgil R. Mahaffey, who works in Central Administration for the

20 Department of Corrections, verified that one of the keys opens the back door of the Dome Building. See

21 Exhibit #354 and Exhibit #232, Pg. 7740.

22   (c) Jayne Gable testified that Frank Gable had an interest in knives, carried

23 knives up his coat sleeve or down the back of his pants, and that she gave him a Chicago cutlery six (6)

24 inch utility knife with a straight tip in December of 1988. See Exhibit #232, Pgs. 7708, 7711, 7712, and

25 7716. Jayne Gable has not seen that particular knife since she gave it to Frank Gable. See Exhibit #232,

26 Pg. 7719. Ms. Gable had two good sets of knives, Chicago Cutlery and Homestead Cutlery, and started

27 missing knives from those sets in January of 1989. She tried to talk Frank Gable into taking the cheap

28 *Frank Edward Gable vs. State of Oregon Post-Conviction Judgment (a:\PCGable121500.JUD)*   Page -21-

Yraguen's Deccision- pages 22 thru 42

1 knives rather than the good ones. See Exhibit #232, Pgs. 7709 and 7710. Mark Gesner also testified

2 about Frank Gable carrying knives; he recalled knives and in particular one which Frank Gable carried

3 in a shoulder sling – Rambo style – which allows the knife to be drawn down and a butterfly knife

4 carried in his coat pocket. See Exhibit #232, Pg. 7981. John Walker also observed Frank Gable with

5 knives – primarily butterfly knives but also one that was a stainless type with a wood or rose wood

6 handle -- being carried in his pocket or coat. See Exhibit #232, Pgs. 8161 and 8162.

7   (i) Mr. Hurley, a forensic evidence expert with the Oregon State

8 Police, testified that after concluding what the class characteristics of the knife, as noted above, used

9 to stab Michael Francke were, he and Mr. Pex, after receiving information regarding the type of Chicago

10 cutlery knife which Janye Gable gave to Frank Gable in December of 1988, purchased the same type of

11 knife. See Exhibit #232, Pgs. 8326, 8349. Mr. Hurley's conclusion was that the Chicago cutlery knife

12 produced results which were consistent with the class characteristics of the knife which was used to stab

13 Michael Francke. See Exhibit #232, Pgs. 8333-8334.

14   (d) Frank Gable started using methamphetamine during the Fall of 1988

15 usually by injection and throughout the period thereafter. Frank Gable quit work at Rigid Truss on

16 November 7th, 1988, the last job he held until he worked for the Marshfield Bargain House in May of

17 1989, and was during the intervening time primarily involved in the use and sale of methamphetamine.

18 Ms. Gable also used methamphetamine. One of Mr. Gable's main associates in selling crank, or

19 methamphetamine, was Mark Gesner. See Exhibit #232, Pgs. 7691 through 7693, 7695 and 7696.

20 Mark Gesner first met Mr. Gable, who was using crank by injection, through John Kevin Walker in June

21 or July of 1988, and Mr. Gable and Mr. Gesner began a relationship of buying and selling drugs See

22 Exhibit #232, Pgs. 7974, 7975, 7979, 7980. John Walker testified that they were all heavy users of

23 drugs, primarily methamphetamine. John Walker did on one occasion observe Frank Gable use a

24 sixteenth (of an ounce) of methamphetamine which would be equivalent in quantity to about two packets

25 of sugar which would be indicative of a high tolerance to the drug. See Exhibit #232, Pgs. 8160 and

26 8161.

27   (e) On one occasion about a month before the murder and during the

28 *Frank Edward Gable vs. State of Oregon Post-Conviction Judgment (a:\PCGable121500.JUD)*   Page -22-

1 period of Frank Gable's heavy use of methamphetamine, Janyne Gable did attempt to access the

2 Department of Corrections computer data, to which she had limited access, at the request of Frank Gable

3 and Mark Gesner, to check on a list of inmate names including Earle Childers, who Frank Gable and

4 Mark Gesner suspected of being an informant. See Exhibit #232, Pg. 7723.

5   (3) Cappie Clifford Harden was in the Salem area during the Fall of 1988 buying

6 and selling cars and selling drugs. See Exhibit #232, Pg. 8056. Cappie Harden first met Frank Gable

7 at a house over on Hyacinth Street at an undetermined time. See Exhibit #232, Pg. 8058.

8   (a) On January 17th, 1989, as he was dropping off Jodie Swearingen,

9 Cappie Harden testified he saw Frank Gable, who was dressed in dark sweats and had a knife in his

10 waistband, parked across the street from the house on Hyacinth Street. See Exhibit #232, Pgs. 8060,

11 8063.

12   (b) Cappie Harden later received two calls from Jodie Swearingen to pick

13 her up at the State Hospital and finally went to the parking lot of the Dome Building between 6:30pm

14 and 7:00pm on January 17th, 1989. See Exhibit #232, Pg. 8064. After Jodie had gotten into his vehicle,

15 a 1970 Mustang, which had to be hotwired in order to start it, Cappie Harden saw a dome light come

16 on in a car across the parking lot and recognized Frank Gable getting into the vehicle. See Exhibit #232,

17 Pgs. 8065, 8066, 8068. Cappie Harden then saw another man who looked like a businessman coming

18 across the parking lot who yelled "Hey, what are you doing in my car?" and started running towards the

19 car. See Exhibit #232, Pg. 8070. Cappie Harden then testified that "Frank came out of the car and stab

20 the man one time in the chest." See Exhibit #232, Pg. 8070. Cappie Harden saw nothing after that

21 because he was busy trying to hot-wire his car and get out of there. See Exhibit #232, Pg. 8070. He

22 took Jodie Swearingen home and "told her to shut up and forget what she ever seen." See Exhibit #232,

23 Pg. 8071.

24   (c) Cappie Harden, when first contacted by police about the homicide on

25 November, 1989, told law enforcement officials that he didn't know what they were talking about,

26 because he was not a rat. However, the police later convinced him that they had evidence that he was

27 lying and he then told them what had occurred. See Exhibit #232, Pg. 8072, 8078. John Storkel cross-

28 *Frank Edward Gable vs. State of Oregon Post-Conviction Judgment (a:\PCGable121500.JUD)*   Page -23-

1 examined Cappie Harden for a considerable period of time during trial (See Exhibit #232, Pgs. 8078

2 through 8145) and established that Mr. Harden had been questioned by the police on November 20th,

3 1989, January 18th, 1990, January 20th, 1990, January 21st, 1990, and January 24th, 1990, giving various

4 accounts and acknowledged lies to the questioning officers during those interviews.

5   (3) Mark McLain Gesner had two contacts with Frank Gable on January 17th,

6 1989, one of which occurred between 8:30pm and 10:30pm. See Exhibit #232, Pg. 7984, 7985, Frank

7 Gable was driving the vehicle owned by Janyne Gable, "a Toyota or Datsun, but it's a Supra". See

8 Exhibit #232, Pg. 7986. Frank Gable came to Mark Gesner's residence and asked him if he would get

9 rid of something for him so no one else could find it and brought in a dark nontransparent plastic bag

10 about two feet wide and two and one-half feet high tied at the top with a knot. When Mark Gesner asked

11 what was in the bag, Frank Gable, who appeared "sweaty" and "nervous" replied "don't worry about

12 it, I'll tell you later. See Exhibit #232, Pgs. 7987, 7988. Mark Gesner had previously gotten rid of

13 methamphetamine laboratory manufacturing materials for Mr. Gable so assumed that that was probably

14 what was in the bag. See Exhibit #232, Pgs. 7982, 7988 Frank Gable was in a hurry and rather than

15 sticking around the way he normally would left immediately. See Exhibit #232, Pg. 7988. The next

16 morning when Mr. Gesner examined the outside of the bag wearing black gloves prior to disposing of

17 it, the bag did not feel like it contained glassware but was rather "spoungy" (sic: spongy) like "something

18 cloth". See Exhibit #232, Pg. 7990. When Mr. Gesner got to the disposal spot he had selected, Sunset

19 Park on the Willamette River, he opened the bag which was tied with a loose knot and got some rocks

20 to put inside the bag. See Exhibit #232, Pg. 7991. As he was holding the bag next to his chest

21 attempting to remove the air from it, "in the middle of the bag I ran into an object that wouldn't bend

22 or – and it felt – it was cylindrical, basically, like something rolled around something hard. See Exhibit

23 #232, Pg. 7990. Mr. Gesner did not examine the contents of the bag but did see a shirt of some type of

24 cloth. See Exhibit #232, Pg. 7992. Mark Gesner threw the bag into the Willamette River in the area

25 looking towards the road next to the fence at the access to the Park. See Exhibit #232, Pg. 7992. Mr.

26 Gesner threw the bag a distance of twenty-five (25) to thirty (30) feet into the water and the bag floated

27 a yard to two (2) yards before disappearing beneath the surface of the water. See Exhibit #232, Pg. 7993.

28 *Frank Edward Gable vs. State of Oregon Post-Conviction Judgment (a:\PCGable121500.JUD)*   Page -24-

1   (a) The Benton County Dive Team was utilized in searching this general

2 area of the Willamette River for a distance of several hundred yards downstream from where the plastic

3 bag was tossed in the River See Exhibit #232, Pg. 8046. Although the Team tried to recreate the

4 situation involving the bag sinking in the manner described by Mark Gesner, the Team could not recreate

5 the same scenario. See Exhibit #232, Pg. 8047-8048. The Willamette River has a depth in that area

6 down to fifteen (15) feet and strong current all the way to the bottom. See Exhibit #232, Pg. 8048.

7 Various man-made items were located on the bottom of the River including a piece of a plastic bag with

8 a rock and some kind of cloth material in it. See Exhibit #232, Pg. 8049, 8050 and 8051. However,

9 upon subsequent examination, the plastic bag contained two pieces of yellow plastic, much like an

10 emergency blanket, and a piece of concrete. See Exhibit #232, Pg. 8251-8252.

11   (4) Keizer Police Officer Kent Barker had first met Frank Gable when he

12 executed a search warrant at the Mark Gesner residence. See Exhibit #232, Pg. 7228. Frank Gable

13 contacted Officer Barker in June of 1989 offering the work for him because of his knowledge of

14 narcotics and weapons dealers in return for keeping him out of jail. See Exhibit #232, Pgs. 7230, 7231.

15 The Keizer Police Department did pay the rent for an apartment for Frank and Janyne Gable for July of

16 1989, but Frank Gable moved out on July 14th, 1989, indicating that he had some things to take care of

17 in Coos Bay and would get back in a couple of days. See Exhibit #232, Pgs. 7246 - 7248.

18   (5) The Gables then lived with David P. Walsh, his fiancee and her children at

19 their apartment located at 1481 Hines in Salem for about a month during August of 1989 before David

20 Walsh took the Gables to Coos Bay. David Walsh had first met Frank Gable through Frank Gable's

21 former Brother-in-Law Randy Studer. See Exhibit #232, Pgs. 7686, 7687, 7930 and 7931. From the

22 time he met Frank Gable near the end of 1988 through January of 1989 and through the time that the

23 Gables lived with him, Frank Gable drove the Toyota Celica owned by the Gables and was heavily

24 involved in using and selling methamphetamine. See Exhibit #232, Pg. 7932.

25   (6) The Oregon State Police received a tip on Frank Gable being recognized from

26 a composite drawing as a possible suspect on June 23rd, 1989, but it was not until August 7th, 1989, that

27 Oregon State Police Detective Darrell J. Berning contacted Frank Gable by telephone and set up an

28 *Frank Edward Gable vs. State of Oregon Post-Conviction Judgment (a:\PCGable121500.JUD)*   Page -25-

1 interview in Salem. However, Frank Gable did not appear for the interview, and he was recontacted

2 and an interview was then set up in Coos Bay for September 13th, 1989. See Exhibit #232, Pgs. 7253,

3 7255 - 7256.

4   b. Verbal statements to law enforcement authorities by Frank Gable:

5   (1) On June 30th, 1989, Keizer Police Officer Kent Barker, who was planning on

6 using Frank Gable for undercover drug work and who had gotten Frank Gable out of jail in Coos Bay,

7 mentioned to Frank Gable on the trip back from Coos Bay that he looked like the individual pictured in

8 the composite drawing of the suspect being sought for the murder of Michael Francke. Frank Gable

9 responded that the Oregon State Police had already talked to him a couple of days after it happened

10 because he was on the grounds the day the murder had happened. See Exhibit #232, Pg. 7235. Frank

11 Gable was wearing an empty knife sheath on his belt. Officer Barker asked him about the sheath, and

12 Frank Gable told Officer Barker that he had lost the knife and then that the Salem Police Department

13 had taken the knife. See Exhibit #232, Pgs. 7236/7237.

14   (2) On September 13th, 1989, when Oregon State Police Officer Jeffrey M.

15 Leighty, who had picked Frank Gable for the interview in the Coos Bay area, was transporting Frank

16 Gable to the interview, Frank Gable volunteered that the reason he was being questioned was because

17 he had been at the Dome Building on the day of the stabbing to pick up his Wife from work. See Exhibit

18 #232, Pg. 7261. Frank Gable also told Officer Leighty that he didn't think that the stabbing took place

19 next to the car because there was no blood. See Exhibit #232, Pg. 7262.

20   (a) When Oregon State Police Officer Dennis R. Fox later questioned

21 Frank Gable on September 13th, 1989, he related that he didn't know where he was, what he was doing

22 or who he was with on the day of the stabbing. Mr. Gable also said that he thought Buck Burgess, an

23 individual who he knew, was a suspect and that he thought that the stabbing was a hit out of the

24 penitentiary. Frank Gable also stated that he wanted to talk further about the investigation with the State

25 Police. See Exhibit #232, Pg. 7271.

26   (3) On September 15th, 1989, Oregon State Police Officer Paul Bain transported

27 Frank Gable from the home of his Stepfather for an interview with Oregon State Police Officer Frederick

28 *Frank Edward Gable vs. State of Oregon Post-Conviction Judgment (a:\PCGable121500.JUD)*   Page -26-

1　E. Ackom. Frank Gable was questioned by Officer Ackom, after being advised of his Mirandi rights,

2　beginning at about 5:00pm on September 15th, 1989. See Exhibit #232, Pg. 7357. During the

3　conversation when Officer Ackom asked Frank Gable what his thoughts were when being questioned

4　about the Michael Francke stabbing, he told Officer Ackom that he was thinking that he had told his

5　Wife in a joking manner that he had killed Michael Francke and that if he had killed Michael Francke

6　it would make him a big man in prison. See Exhibit #232, Pg. 7292/7293. At one point in the interview,

7　Frank Gable began crying and stated that "I just know they're going to roast me or fry me for this." See

8　Exhibit #232, Pgs. 7294/7352. Frank Gable also related that he knew that the knife that the Salem Police

9　Department had was not the knife which was used in the stabbing, and that he thought that the knife used

10　in the stabbing would have been gotten rid of outside of the Salem area. See Exhibit #232, Pg.

11　7295/7296. He also stated that Chris had given him a knife a couple of weeks before the murder, and

12　that on the night of the murder he had to be with Chris. See Exhibit #232, Pg. 7349. He later said that

13　on the night of the murder he was with Chris (Warilla) or at home with his Wife and friends, whom he

14　identified as Jason Farm, Kevin Walker and a guy named Russell. See Exhibit #232, Pg. 7350. After

15　being questioned by Officers Bain and Berning, as noted below, Frank Gable was recontacted by Officer

16　Ackom at about 1:20am on September 16th, 1989. See Exhibit #232, Pg. 7357. Frank Gable told

17　Officer Ackom that he wasn't on the State Hospital grounds when Michael Francke was killed, did not

18　kill Michael Francke and didn't know who killed him, and that "I know that's going to sound pretty

19　stupid on the stand, but that's the story I'm sticking with" to the end. See Exhibit #232, Pgs.

20　7350/7351/7352. He also said that the back part of his brain was telling him "you did it" but all the time

21　he knew he didn't do it. See Exhibit #232, Pg. 7352. He also said "Yeah, but I'm not saying I did it.

22　I'll go to the end of the trial saying that, Fred. There are only two people who know who killed Francke

23　and God." When questioned about that statement since Michael Francke was dead, he said "Well, then

24　there are only two people who know Francke – yeah, me and God." And after he realized what he had

25　said, again said "Yeah, Yeah. Me and God." See Exhibit #232, Pgs. 7353/7354. When Officer Ackom

26　requested permission to search his Mother-in-Law's residence, at Oregon State Police Sergeant Salle's

27　request, Frank Gable stopped the conversation and requested a lawyer. The conversation ended at that

28　*Frank Edward Gable vs. State of Oregon Post-Conviction Judgment (a:\PCGable121500.JUD)*　　　　Page -27-

506

1 point. The time was about 3:10am on September 16th, 1989. See Exhibit #232, Pg. 7357.

2 (a) About 9:00pm between conversations between Frank Gable and Officer

3 Ackom on September 15th, 1989, Oregon State Police Officer Paul Bain, accompanied by Officer Darrell

4 J. Berning, questioned Frank Gable after advising Frank Gable of his Miranda rights. Frank Gable

5 related that his Wife Jayne had worked at the Oregon State Hospital on the day of the homicide, and that

6 he had picked her up at about 3:30pm. See Exhibit #232, Pgs. 7308/7309/7327. He further stated that

7 he had stayed with his Wife all evening, and that there had been a crank party at their apartment

8 involving about 20 to 30 people. See Exhibit #232, Pg. 7309. Frank Gable was unable to recall any

9 names of the individuals attending. See Exhibit #232, Pg. 7320. Frank Gable pointed to the area on a

10 aerial photograph where Michael Francke parked, and said he knew where Michael Francke parked

11 because he had worked at the Dome Building and had picked up garbage in the parking area. See

12 Exhibit #232, Pg. 7324. Frank Gable also reported that he had driven his Wife Jayne to work at Building

13 34 on the Oregon State Hospital grounds at about 6:30am on January 18th, 1989. See Exhibit #232, Pg.

14 7327. At the conclusion of that interview, Frank Gable was arrested on a parole and probation detainer

15 which involved an alleged assault on his Wife Jayne. See Exhibit #232, Pg. 7317.

16 (i) Janyne Gable testified at trial that Frank Gable picked

17 her up at noon on January 17th, 1989, because she had become ill. Frank Gable took Janyne Gable home,

18 and the last time she saw Frank Gable that day was when he left at about 1:00pm. Janyne Gable did not

19 see Frank Gable until about 6:25am, about which she was disgusted, the morning of January 18th, 1989,

20 when he came by to take her to work at 6:30am. Janyne Gable never saw Frank Gable the evening or

21 night of January 17th, 1989, and Frank Gable had left with their Toyota Celica and returned the next

22 morning with the same vehicle. See Exhibit #232, Pgs. 7727 through 7729. Their normal route to and

23 from work was along D Street. See Exhibit #232, Pg. 7730.

24 (b) During the transport at about 11:00pm on September 15th, 1989, to the

25 Coos County Jail in Coquille, Frank Gable volunteered to the transporting Oregon State Police Officer

26 Michael Wm. Stupfel that he wasn't even near the Dome Building when the guy was killed and that he

27 would kill himself. See Exhibit #232, Pgs. 7337/7338. Officer Ackom recontacted Frank Gable on

28 *Frank Edward Gable vs. State of Oregon Post-Conviction Judgment (a:\PCGable121500.JUD)*   Page -28-

1   September 16th, 1989, about his statements to Officer Stepfel, and Frank Gable told Officer Ackom that

2   it didn't matter whether he said "yes" or "no", he knew that they were going to roast him for the murder

3   (See Exhibit #232, Pgs. 7346/7347), and that Officer Ackom didn't need a suicide watch, because he

4   wasn't going to kill himself.

5   (4) At approximately 8:00pm on September 16th, 1989, after being served with

6   a search warrant for hair and blood samples, Frank Gable was transported to the Coquille Valley

7   Hospital by Oregon State Police Officer David M. Gifford accompanied by Sergeant Thomas E. Benz.

8   See Exhibit #232, Pg. 7364. Officer Gifford and Frank Gable were outside the emergency room, when

9   they both overheard Dr. Count say that he wanted to follow the procedure closely or some attorney

10   would get the guy off. Mr. Gable who was present responded "No lawyer in the world could get me

11   off." See Exhibit #232, Pgs. 7365/7366. Dr. Count later said the guy was up on a murder charge, and

12   Mr. Gable responded "I wasn't fucking arrested for murder", turned pale and began uncontrollably

13   crying. See Exhibit #232, Pgs. 7366/7371.

14   (5) On or about November 3rd, 1989, Frank Gable requested Coos County

15   Corrections Officer John S. Newcomer to contact Oregon State Police Officer Paul Bain, because he

16   wanted to speak with him about the Francke case. See Exhibit #232, Pgs. 7453 - 7455. Frank Gable

17   also called the Oregon State Police Headquarters requesting to speak to Officer Bain or any other

18   detectives who had previously talked to him. See Exhibit #232, Pg. 7457. Officer Bain, Officer

19   Ackom and Sergeant Salle responded and met with Frank Gable at the Coos County Jail in Coquille

20   approximately 4:45pm on November 3rd, 1989 concluding the conversation at 9:58pm. See Exhibit

21   #232, Pgs. 7458/ Frank Gable was again given his Mirandi rights and was advised that the conversation

22   was being recorded. See Exhibit #232, Pgs. 7458, 7459, 7485; see also Exhibit #206 which is a

23   transcript of the conversation.

24   (a) Frank Gable mentioned an article in the newspaper quoting Mike

25   Keerins as saying that Frank Gable had confessed to him that he had killed Michael Francke. See

26   Exhibit #232, Pg. 7460. Mr. Gable advised the Officers that Mike Keerins was lying and that he did not

27   make any confession to him. See Exhibit #232, Pg. 4761. Later, Frank Gable related he believed that

28   *Frank Edward Gable vs. State of Oregon Post-Conviction Judgment (a:\PCGable121500.JUD)*   Page -29-

1   the only person making such statements to the police was his Wife Jayne, and that he didn't believe that

2   the police had other people making such statements. See Exhibit #232, Pg. 7512. Frank Gable was told

3   by Sergeant Salle that five people were saying that he was involved in the murder, to wit, Chris Warilla,

4   Gayla Freeman, Mark Gesner, Kris Keerins and Jodie Swearingen. See Exhibit #232, Pg. 7548.

5   (b) Mr. Gable advised that he had picked up his Wife Jayne at about

6   3:30pm, although he acknowledged that it might have been at about 12:30 noon instead on January 17th,

7   1989, and that he had been with Chris Warilla the night of January 17th, 1989, selling some

8   methamphetamine at an AM/PM store at about 7:30 to 8:00pm and then cruised around Salem, although

9   he later said he couldn't be sure where he was that night. See Exhibit #232, Pgs. 7464 -

10 7466/7476/7479/7488. He also said that he stayed the whole night at Chris' house and returned to his

11 house the next morning and took his Wife Jayne to work. See Exhibit #232, Pg. 7465. However, in

12 summary, during the conversation, Frank Gable told Officer Bain that he was a Chris Warilla's house,

13 that he was at Shelli Thomas' house, that he was at his house, or that he could have been on a rocket ship

14 ride to the moon and back at the time that Michael Francke was murdered, but he was sure he wasn't

15 on the State Hospital grounds and didn't kill Michael Francke. See Exhibit #232, Pgs. 7522/7524.

16   (c) Frank Gable acknowledged that the knew Randy Studer, Earle

17 Childers, Jodie Swearingen, Mark Gesner, and Johnny Crouse. See Exhibit #232, Pgs. 7466 -

18 7470/7478. Mr. Gable advised that he knew the police had no physical evidence, and that there would

19 never be any physical evidence. See Exhibit #232, Pg. 7489. Frank Gable advised the Officers that they

20 didn't have anything other than circumstantial evidence, and that God had directed him to help the

21 Oregon State Police put their investigation in the right direction. See Exhibit #232, Pgs. 7498/7558.

22 Frank Gable told the Officers that he would look at pictures of people they were questioning and if they

23 would tell him what the people were saying about him, he would tell the police what he knew about

24 them. He knew there were people saying bad things about him and that he might have to eat this thing.

25 See Exhibit #232, Pg. 7490. Frank Gable concluded the conversation with the Officers at about 9:40pm

26 on November 3rd, 1989.

27   (6) Officer Frederick E. Ackom again had a conversation with Frank Gable, after

28 *Frank Edward Gable vs. State of Oregon Post-Conviction Judgment (a:\PCGable121500.JUD)*   Page -30-

1 advising him of his Mirandi rights, for about two hours beginning at about 9:30am on December 22nd

2 1989, at the Coos County Jail in Coquille. See Exhibit #232, Pgs. 7561 - 7563. Frank Gable was given

3 a list of names and the only name he reacted to was Jodie Swearingen, and as Officer Ackom was giving

4 Mr. Gable a second list of names Frank Gable said "That Jodie gal, the bitch is saying she saw me run

5 from the scene, isn't she?". See Exhibit #232, Pgs. 7566/7575/7590. Frank Gable also acknowledged

6 that he could have been at John and Kelly Bender's place on Hyacinth Street in Salem the night that

7 Michael Francke was killed, because he was doing dope deals at that house. See Exhibit #232, Pg. 7567.

8 He also said without being asked that if anyone had seen him the night that Michael Francke was

9 murdered, he would have been wearing dark sunglasses. See Exhibit #232, Pgs. 7569 - 7571. Even

10 though Officer Ackom was ill, he stopped by that same afternoon at about 2:30pm and continued his

11 conversation with Frank Gable for about two hours. See Exhibit #232, Pg. 7574.

12   (7) Officer Ackom again had a conversation with Frank Gable, after advising him

13 of his Mirandi rights, on January 21st, 1990, beginning at about 6:00pm at the Coos County Jail, which

14 only lasted about four and one-half minutes. See Exhibit #232, Pg. 7577 - 7578. Frank Gable reiterated

15 that all they had was a bunch of zeros – a bunch of circumstantial evidence. See Exhibit #232, Pg. 7577.

16 Officer Ackom told Frank Gable that they now had two people who said they saw him run from the

17 scene of the murder. Frank Gable became agitated, left the room and told Officer Ackom that he didn't

18 want to hear about it. See Exhibit #232, Pg. 7579.

19   (8) On April 8th, 1990, at approximately 5:09am Frank Gable was arrested by

20 Oregon State Police Officer Loren T. Glover at the Coos County Jail on a Warrant issued on an

21 Indictment charging Frank Gable with six (6) Counts of Aggravated Murder and one (1) Count of

22 Murder for the stabbing of Michael Francke. See Exhibit #232, Pgs. 7594/7595. Frank Gable was then

23 taken in a van with two other unmarked vehicles to the State Police Office in Florence arriving about

24 7:45 am. See Exhibit #232, Pgs. 7598/7621.

25   (a) During the stop in Florence Frank Gable was questioned by Officers

26 Glover and Ackom. Officer Glover had had some limited involvement in the investigation and at one

27 point said to Mr. Gable: "Frank, do you realize that we have a witness that can place you at the scene?

28 *Frank Edward Gable vs. State of Oregon Post-Conviction Judgment (a:\PCGable121500.JUD)*   Page -31-

1   That saw you drive away from the scene on to D street and then head up to Park Street and turn left?"

2   Frank Gable replied: "Oh, Earle told you that, didn't he." Officer Glover replied: "Oh, how did you

3   know it was Earle?" Frank Gable said "Well, I knew that Earle worked – his wife worked over there

4   at the hospital." See Exhibit #232, Pg. 7604. When Officer Glover was interviewing Mr. Gable about

5   being at Mark's (Gesner) house with Rachel when Earle showed up after walking away from the

6   Corrections Division Release Center, all of which Frank Gable remembered, Officer Glover asked Mr.

7   Gable, "Well, on this walk, did you tell Earle that you killed Michael Francke?" Frank Gable replied:

8   "No, I didn't tell him that." See Exhibit #232, Pg. 7607. Officer Ackom at one point told Mr. Gable

9   that he was going to get convicted on this crime, and Frank Gable said "Well, I'm content I can do the

10   time, makes no difference to me. Doesn't matter whether I'm guilty or not, I can do the time, makes no

11   difference." See Exhibit #232, Pg. 7610. Frank Gable continued to deny any responsibility for the

12   murder of Michael Francke.

13   (b) After several hours, to wit, until about 3:30pm, the party left Florence

14   and traveled to the Springfield State Police Office. See Exhibit #232, Pgs. 7611 - 7612. Frank Gable

15   continued to deny any responsibility for the murder of Michael Francke, but at one point after Frank

16   Gable told the Officers that they had the wrong guy and that the killer was still on the streets, Oregon

17   State Police Sergeant McCafferty said: "No, Frank, we got the killer sitting right here." Mr. Gable

18   replied: "Maybe so, maybe not". Officer Glover said: "You don't have the guts to tell us about killing

19   Michael Francke. You're going to take this to the grave with you, aren't you?" Mr. Gable replied "You

20   bet I am." When being asked by Officer Glover why out of all the names given to him he picked out

21   Earle [Childers] as being the one who saw him driving away from the Dome Building, Mr. Gable

22   responded "Well, just luck, I guess." See Exhibit #232, Pg. 7613. Mr. Gable was lodged in the Lane

23   County Jail and the next morning was taken to the Marion County Jail in Salem. See Exhibit #232, Pg.

24   7614.

25   (9) On April 9th, 1990, just prior to Frank Gable's arraignment, Officer Ackom

26   had a brief conversation with him in the Marion County Jail. Mr. Gable was shaking his head from side

27   to side and told Officer Ackom: "You have got the wrong guy, Fred. I don't know why these people are

28   *Frank Edward Gable vs. State of Oregon Post-Conviction Judgment (a:\PCGable121500.JUD)*   Page -32-

1 saying these things about me. I wish I could tell you some things about them. I wouldn't do that. I'm

2 not a rat. The best I can do for you, Fred, is I might have been driving by that night and Jodie and Shorty

3 saw me." See Exhibit #232, Pg. 7630.

4   c. Verbal statements made to individuals, other than law enforcement officials, by Frank

5 Gable:

6   (1) Linda M. Perkins, who was the Mother of Theresa Ross, who from December

7 of 1988 through February of 1989 was living with Randy Studer, who at the time was Frank Gable's

8 Brother-in-Law, recalls seeing Frank Gable arrive in his Toyota vehicle at the Studer residence during

9 the early morning after Michael Francke had been killed. Frank Gable was nervous and agitated and told

10 Theresa Ross that "I fucked up.", "I fucked up big time this time." Linda Perkins said "What do you

11 mean?". Frank Gable said "Well, I'll put it to you like this; you will be reading about it in the papers."

12 See Exhibit #232, Pgs. 7952, 7969. About a week later in a telephone conversation, Frank Gable told

13 Linda Perkins that if she ever said anything, she would be a dead fucker. See Exhibit #232, Pg. 7953.

14   (a) On cross-examination it was developed that Linda Perkins had

15 contacted the Francke Task Force and was interviewed on October 8th, 1989; that she was upset at the

16 time of the conversation with Frank Gable because she knew that he was supplying both her Daughter

17 and Randy Studer with drugs; and that when she asked her Daughter about what Frank Gable had said,

18 she just said that Frank was "full of bullshit". See Exhibit #232, Pgs. 7967-7968. It was also during this

19 period of time that Linda Perkins had seen Jodie Swearingen riding around in Frank Gable's Toyota

20 vehicle. See Exhibit #232, Pg. 7968.

21   (2) John Kevin Walker met Frank Gable in about September or October of 1988

22 through Russ Estep who was living in the same apartment complex as Frank Gable on Hawthorne. Russ

23 Estep was supplying drugs to Frank Gable which Mr. Estep was obtaining from Mr. Walker. Later, John

24 Walker began directly supplying Frank Gable with methamphetamine. See Exhibit #232, Pgs. 8153

25 through 8158. On January 17th, 1989, John Walker returned a telephone call to Frank Gable, and Frank

26 Gable wanted him to come into town early. John Walker was attending his Brother's birthday party first

27 and told Mr. Gable that he wouldn't be in town until about 8:00pm to 9:00pm that evening. See Exhibit

28 *Frank Edward Gable vs. State of Oregon Post-Conviction Judgment (a:\PCGable121500.JUD)*   Page -33-

1   #232, Pg. 8165. Later that night when he was in Salem during the early morning hours of January 18th,

2   1989, he heard on his scanner that something was going down at the State Hospital and drove by that

3   area observing that there were lots of police cars in the area of the Dome Building. Mr. Walker went

4   to Paul Farm's house and took the portable portion of his scanner into the house and listened to the

5   police activity until about 3:00am before he left to return to Corvallis. See Exhibit #232, Pgs. 8167

6   through 8169. Mr. Walker worked the next day and received two telephone calls from Frank Gable who

7   appeared to be in a state of "real tweaked agitation", and was edgy and jittery like he needed another fix.

8   See Exhibit #232, Pg. 8171. Mr. Walker finally saw Mr. Gable about 8:00pm to 9:00pm on January 18th,

9   1989, in Salem at Mr. Gable's house. When they went into the bedroom for the exchange of drugs, Mr.

10   Gable asked if Mr. Walker had heard the news and Mr. Walker asked what news? Mr. Gable said:

11   "About that guy over there at the State Hospital grounds." And Mr. Walker said: "Yeah, he got shot or

12   something." Mr. Gable replied: "Well, that's not exactly what happened, but it's close enough. I stuck

13   him." Mr. Walker testified that Mr. Gable's attitude seemed remorseful and as soon as Mr. Gable

14   realized what he had said, he told Mr. Walker, while he was handling the .357 previously sold to him

15   a couple of weeks before by Mr. Walker which firearm had been laying on the table, "don't tell on me,

16   Kevin, or I'll have to kill you and kill your family." See Exhibit #232, Pg. 8173. Mr. Walker didn't see

17   Mr. Gable as often after that but observed that Mr. Gable was real edgy, like a tweaker's edge, involved

18   more and more in the drug scene. See Exhibit #232, Pg. 8177.

19   (a) Police first tried to interview John Walker in September of 1989, but

20   he refused to cooperate because he was afraid and because he didn't want a rat jacket. See Exhibit #232,

21   Pg. 8179. Mr. Walker later agreed to cooperate in exchange for an agreement from the State not to

22   prosecute him for any non-person crimes revealed. See Exhibit #232, Pg. 8184. John Walker suspected

23   that Frank Gable was a snitch and when he and Mark Gesner later in 1989 saw Frank and Janyne Gable

24   appear at Mark Gesner's house, they knew they were going to get busted. See Exhibit #232, Pg. 8200.

25   (3) A day or two after Frank Gable had asked Mark Gesner to get rid of the black

26   plastic bag described in notes above (See Exhibit #232, Pgs. 7985 through 7994) given to Mr. Gesner

27   by Mr. Gable the night of January 17th, 1989, Frank Gable called Mark Gesner and asked him if he had

28   *Frank Edward Gable vs. State of Oregon Post-Conviction Judgment (a:\PCGable121500.JUD)*   Page -34-

1   gotten rid of the bag for him. Mr. Gesner told Mr. Gable that he had disposed of the bag.

2   (a) After disposing of the bag, the relationship between Mr. Gesner and

3   Mr. Gable continued as before. See Exhibit #232, Pg. 7996. Mark Gesner did not reveal the information

4   he had during the first contacts by the Oregon State Police, because he was incarcerated and feared a

5   snitch jacket. See Exhibit #232, Pg. 8000. Mr. Gesner finally told the State Police on February 24th,

6   1990, about the bag after discussing the matter with his attorney, Paul Ferder, and assuming that the

7   State Police knew he wasn't telling the truth and wanting to avoid any potential for a hindering

8   prosecution charge. See Exhibit #232, Pgs. 8001, 8002 and 8015. The State did agree that if there was

9   a basis for a Hindering Prosecution charge that they would not pursue such, the Parole Board was

10   advised that Mr. Gesner was cooperating in the Francke investigation, and the State Police assisted in

11   getting him removed from segregation at the Federal Sheridan facility. See Exhibit #232, Pgs. 8004

12   through 8006. Mark Gesner did have knowledge that Frank Gable had acted as a snitch for the Keizer

13   Police Department at the time that he revealed his information about the Francke murder and felt that

14   Frank Gable had probably snitched on him before the drug raid on his residence at 714 Pine Street,

15   although he later learned that John Larimore was the person advising the police. See Exhibit #232, Pg.

16   8021 and 8022.

17   (4) A day or two after Earle Childers had observed Frank Gable in his Toyota

18   pulling out from the medical building across the street from the Dome Building between 6:30pm and

19   7:00pm on January 17th, 1989, as more fully set forth below, Earle Childers had a conversation with

20   Frank Gable about having seen him the evening of January 17th. Frank Gable told Earle Childers to

21   forget that he had ever seen him at that location. See Exhibit #232, Pg. 7757.

22   (a) On one later occasion, when Frank Gable was speaking about Shelli

23   Thomas' boyfriend, who had been mistreating her, he told Earle Childers that he would just stick the

24   boyfriend, and that it "won't be the first time." See Exhibit #232, Pg. 7759.

25   (b) Later, on or about July 11th, 1989, after Earle Childers had absented

26   himself from DCRC, a Department of Corrections facility, and gone to Mark Gesner's house. Frank

27   Gable was there and told them that he had been picked up and questioned about Michael Francke's

28   *Frank Edward Gable vs. State of Oregon Post-Conviction Judgment (a:\PCGable121500.JUD)*   Page -35-

1 death, but they had let him go because they had nothing on him. See Exhibit #232, Pg. 7764. When

2 Frank Gable was asked if he did it, Frank Gable just smiled and insinuated more yes than no. Later that

3 same evening when Earle Childers was walking with Frank Gable to Mr. Gable's residence to get some

4 clothes, Frank Gable said that he had done it; that he had been burglarizing the car, going through some

5 cars and was in Francke's car and got caught and ended up sticking him because he didn't want to go

6 back to prison. Frank Gable said that he had stuck him three or four times in the chest. See Exhibit

7 #232, Pg. 7767. Frank Gable said that he was going through the unlocked Francke car trying to find a

8 gun; that Michael Francke was a cock sucker and now he would always be a cock sucker. See Exhibit

9 #232, Pg. 7768.

10   (c) Earle Childers first told the law enforcement authorities on September

11 20th, 1989, that he knew nothing about the Michael Francke murder, but on subsequent questioning on

12 November 22nd, 1989, revealed the above information. See Exhibit #232, Pg.7771.

13   (5) David P. Walsh had a conversation in February or March of 1989 when he was

14 throwing a knife across from where Shelli Thomas lived, prior to David Walsh moving to a different

15 residence and prior the Gables moving in with him. David Walsh told Mr. Gable that he had gotten the

16 knife from Jerry Paul Baker. Frank Gable, who was under the influence of drugs, said that he had given

17 the knife to Baker and that it was the knife that he had used to kill Michael Francke. See Exhibit #232,

18 Pg. 7934. Frank Gable also told David Walsh if he ever said anything that he would kill him and his

19 family. See Exhibit #232, Pg. 7935.

20   (a) David Walsh said that he later pawned the knife. See Exhibit #232,

21 Pg. 7941.

22   (b) Later, after the Gables had moved in with David Walsh, Frank Gable

23 made a comment about killing Doug Scratchfield. Frank Gable said "Well, you remember Michael

24 Francke" and said that that would happen to Doug Scratchfield. See Exhibit #232, Pg. 7936. Frank

25 Gable further said the "he had been jockey-boxing and the car had the car door open, that the he was

26 laying across the seat and that Mr. Francke came up along to me up on him, and he had lunged out into

27 Mr. Francke" stabbing him repeatedly and fled across the parking lot. Frank Gable again threatened

28 *Frank Edward Gable vs. State of Oregon Post-Conviction Judgment (a:\PCGable121500.JUD)*   Page -36-

1 David Walsh and his family if he said anything. See Exhibit #232, Pg. 7937.

2   d. Evidence placing Frank Gable at or near the scene of the murder of Michael Francke

3   (1) During late 1988 Earle Francis Childers had met Frank Gable through Mark

4 Gesner and began to give Dilaudids to Frank Gable in exchange for crank (methamphetamine). The

5 methamphetamine would be obtained by Frank Gable from a house on Hyacinth Street but Earle

6 Childers never went inside the residence. See Exhibit #232, Pgs. 7743, 7746 and 7747. Earle Childers

7 was aware that Frank Gable was carrying knives around this period and can recall in particular a brown

8 handled knife like a hunting knife. See Exhibit 232, Pgs. 7747 and 7748. Earle Childers was on parole

9 and was required to attend NA/AA meetings at Welcomaa Club in the Lancaster Mall area. The NA/AA

10 meetings were held at 5:00pm and were over at 6:30pm. Earle Childers, who was walking to and from

11 the meetings, would pass by the State Hospital at about 7:00pm. On the night of January 17th, 1989,

12 between 6:30pm and 7:00pm, he was coming down D Street and saw Frank Gable pull out "by the

13 medical building, come up, turn right and start up D Street." See Exhibit #232, Pgs. 7754 and 7755.

14 Frank Gable was wearing sun glasses, and Earl Childers tried to hail him but Frank Gable did not stop.

15 The Gable vehicle proceeded up D Street to Park Street and turned left and was not seen again that night

16 by Earle Childers. See Exhibit #232, Pgs. 7754-7756.

17   e. No trace or transfer evidence was discovered by investigating authorities linking Frank

18 Gable with the murder of Michael Francke. See Exhibit #232, Pg. 8407.

19

20

21 ***III. DEFENDANT'S CASE IN CHIEF***

22   1. Matters relating to the credibility of State's witness Janyne Gable:

23   a. Lynn Ellen Studer, Janyne Gable's Mother, testified that Janyne Gable "frequently told

24 lies and manipulated me and other people in situations." See Exhibit #232, Pg. 8802. Mrs. Studer

25 stated that the marriage between Frank and Janyne Gable was deteriorating, Frank and Janyne Gable

26 were using drugs, and Frank and Janyne Gable had an "interest in a drug related capacity" with her Son

27 Randy Studer. See Exhibit #232, Pgs. 8803, 8804. Mrs. Studer didn't want her Daughter to associate

28 *Frank Edward Gable vs. State of Oregon Post-Conviction Judgment (a:\PCGable121500.JUD)*   Page -37-

1 with Frank Gable because of the drugs and violent living situation. See Exhibit #232, Pg. 8805.

2   b. Kristin Kathleen Studer, Janyne Gable's Sister, stated that Janyne Gable "stretched the

3 truth", was "vague", manipulated and lied. See Exhibit #232, Pg. 8812. Kristin Studer knew that Frank

4 and Janyne Gable were involved in drugs and did not feel that the relationship between Frank and Janyne

5 Gable was good for either one of them. See Exhibit #232, Pgs. 8812, 8813.

6   2. Matters relating to the credibility of State's witness David Walsh:

7   a. Sheryle Lowery, the former girlfriend of David Walsh, testified that she left him

8 because of his involvement in drugs. See Exhibit #232, Pg. 8479. Ms. Lowery also acknowledged that

9 Frank and Janyne Gable moved in with them for about a month during the Summer of 1989 and moved

10 out on friendly terms because the Gables had to go to court in Coos Bay. Ms. Lowery stated that none

11 of them had jobs during the period they were together, and that Frank Gable never mentioned Michael

12 Francke when she was around. See Exhibit #232, Pgs. 8480 through 8482.

13   b. Jerry Paul Baker, who was incarcerated at the time of testifying and acknowledged

14 previous convictions, testified that he knew David Walsh because they sold dope back and forth. See

15 Exhibit #232, Pg. 8579. During the Spring or Summer of 1989 Jerry Baker traded a car stereo to Frank

16 Gable for a G.I. type hunting knife which was about 1 3/4 inches to 2 inches wide and an eighth of an

17 inch thick. See Exhibit #232, Pgs. 8576 through 8578. He met Frank Gable through Shelli Thomas.

18 See Exhibit #232, Pg. 8579.

19   3. Matters relating to the credibility of State's witness Mark Gesner:

20   a. Pamela Renee Winn testified that she was married to Mark Gesner, who she said was

21 a habitual liar and a person whom one could not believe a word he said. See Exhibit #232, Pgs. 8491-

22 8492. She thought that Mark Gesner was a closer friend of Earle Childers than of Frank Gable. See

23 Exhibit #232, Pg. 8495. Ms. Winn also knew that Frank Gable was into drugs and "prowling around",

24 to wit, ripping things off. See Exhibit #232, Pgs. 8497-8498.

25   b. Philip A. Stanford, a columnist for the Oregonian, had contact with Mark Gesner

26 during the Fall of 1989 when Mr. Gesner was in the Clark County Jail. See Exhibit #232, Pgs. 8505 -

27 8506. Mr. Stanford acknowledged those details which appeared in his column, but declined to discuss

28 *Frank Edward Gable vs. State of Oregon Post-Conviction Judgment (a:\PCGable121500.JUD)*   Page -38-

1 or release any other information from the taped conversation with Mr. Gesner based upon privilege. Mr.

2 Stanford reported in his column that Mr. Gesner told him that Frank Gable hadn't told him anything, and

3 that the authorities were willing to make a deal with him but he wasn't willing to lie to get someone the

4 gas chamber. See Exhibit #232, Pgs. 8507 through 8513.

5   4. Matters relating to the credibility of State's witness John Walker:

6   a. Mark Dean Davis, an incarcerated inmate at the Oregon State Penitentiary who

7 acknowledged the various crimes for which he had been convicted, said he had contact with John Walker

8 sometime during June of 1990 when both were in custody and transferred on a belly chain to the Marion

9 County Courthouse. See Exhibit #232, Pgs. 8521-8522. Mark Davis asked John Walker about being

10 him being hit in the face, and John Walker told him that he wasn't going to testify against Frank Gable;

11 that the District Attorney was dismissing some charges but when he came to court he was going to plead

12 some kind of amendment. Mark Davis then told John Walker that he didn't care what happened to

13 Frank Gable, and John Walker said he was doing this because of some kind of drug business he and

14 Frank had on the street and that he was mad at Frank Gable. See Exhibit #232, Pg. 8523.

15   b. Peter Mitchell Baker, an incarcerated inmate who acknowledged the various crimes

16 for which he had been convicted, said he had sought out John Walker at the Oregon State Correctional

17 Institution in 1990 and had an altercation with John Walker in which he more or less had hit him in the

18 face because he said something unpleasant about his girlfriend Shelli Thomas. See Exhibit #232, Pgs.

19 8531, 8532 and 8535. On June 21st, 1989, he told Oregon State Trooper Pileggi that he hoped that they

20 never found who had killed Michael Francke. Exhibit #232, Pg. 8537. Later, he told a defense

21 investigator that "I hate rats. Keven Walker was a rat. If I saw Shorty Harden, I would smash his face

22 because I hate rats." See Exhibit #232, Pg. 8538.

23   (1) Shelli Lynn Thomas testified that Peter Baker had written her a love letter

24 from prison in which he wrote that he had broken Mr. Walker's face. See Exhibit #232, Pgs. 9393, 9395

25 & Exhibit #619 at Trial.

26   c. Dennis Dean Gause, an incarcerated inmate, said that he saw John Walker in jail

27 sometime after September of 1989 and that John Walker said he didn't know anything about the Michael

28 *Frank Edward Gable vs. State of Oregon Post-Conviction Judgment (a:\PCGable121500.JUD)*   Page -39-

518

1 Francke murder and just wanted to be left alone. See Exhibit #232, Pg. 8864.

2   5. Matters relating to the credibility of Cappie Harden aka "Shorty" Harden:

3   a. Larry Seeley, a Deputy Sheriff with the Marion County Sheriff's Office, transported

4 Cappie Harden for surgery on February 23rd, 1990, and was with Cappie Harden after surgery as he was

5 regaining consciousness. Cappie Harden was making a few rambling incoherent comments among

6 which he said: "I'm out of here as soon as I hang a murder rap on Frank Gable." See Exhibit #232, Pgs.

7 8543 through 8547.

8   b. Patrick Ray Boggs, an individual who acknowledged some former convictions, said

9 that Cappie Harden owed him $350, which he still had not collected, and when he saw Cappie's vehicle

10 at Sam Harmon's house at 3929 Iberis Street in Salem in late April, 1990, he stopped and had a

11 conversation with Cappie Harden. See Exhibit #232, Pgs. 8564, 8568 and 8571. Cappie Harden said

12 that with respect to his testimony against Frank Gable, he was only doing what they told him to do

13 because his ass was on the line, that he stood to make money out of the deal, and that he didn't know

14 what the hell went on with respect to the murder of Michael Francke. See Exhibit #232, Pg. 8566.

15 Patrick Boggs believes that Mr. Harden set him up on matters for which he was arrested. See Exhibit

16 #232, Pg. 8570.

17   c. Dwayne Elmo Christiansen, who was incarcerated and acknowledged previous

18 convictions, when he was moved from segregation to the same pod occupied by Cappie Harden during

19 June of 1990 and who had not met Mr. Harden until that time, asked where Cappie Harden was. When

20 he found him he asked why he was doing this to Frank? Cappie Harden told him that he was simply

21 telling on Frank before Frank could tell on him about things he had done on the streets. Mr. Harden

22 further told him that he was a million dollar baby, that he was going to receive money for telling on

23 Frank, and that they were trying to get Frank out of the way. See Exhibit #232, Pg. 8583.

24   d. Adam Manuel Hernandez, who was incarcerated and acknowledged previous

25 convictions, testified that he knew Shorty Harden because he on worked on cars for him. Mr. Hernandez

26 said that he did not run errands for Mr. Harden, did not give Jodie Swearingen a ride from Dundee to

27 Salem on January 17th, 1989, and that Mr. Harden was a liar and a fat mouth. See Exhibit #232, Pgs.

28 *Frank Edward Gable vs. State of Oregon Post-Conviction Judgment (a:\PCGable121500.JUD)*   Page -40-

1   8594 through 8596.

2   e. Kenneth John Beeler, who acknowledged previous convictions and was incarcerated

3 in a Marion County Jail pod with Cappie Harden in February of 1990, spent five days in the hole with

4 Cappie Harden. Mr. Beeler said that Mr. Harden changed his story from minute to minute about the

5 Frank Gable, and that Cappie Harden said he had five or six or seven cases against him so he had to help

6 the State on the Francke case. Mr. Harden also said that he was a million dollar baby, was going to get

7 part of the reward money, and movie royalties for his testimony but that he went back and forth on

8 whether he and/or Jodie Swearingen were or were not at the murder scene. See Exhibit #232, Pgs. 8614

9 through 8617.

10   f. Dennis Dean Gause, who was incarcerated with Cappie Harden in the hole in the

11 Marion County Jail for a couple of months beginning in about December of 1989, said that Mr. Harden's

12 first story was that he gave Frank Gable and Jodie Swearingen a ride to the Dome Building on January

13 17th, 1989; then Mr. Harden failed the polygraph and he said that he got a call from Jodie Swearingen

14 and went to the Dome Building, saw what happened and gave Frank Gable and Jodie Swearingen a ride

15 from the Dome Building. See Exhibit #232, Pgs. 8849 through 8851. Mr. Gause also said that he, Bill

16 Storm and Cappie Harden were using the whole thing as a scam to get out of jail, but that Cappie Harden

17 was the only one whom the police believed. See Exhibit #232, Pgs. 8860 through 8862. Mr. Gause also

18 testified that Cappie Harden and Mike Keerins said they would get cash for testifying and that Cappie

19 Harden referred to himself as a million dollar baby. See Exhibit #232, Pgs. 8864, 8865.

20   g. John Allen Bender and his Wife, Kelly Ann Bender, testified that they met Frank Gable

21 at Circle K for a drug transaction three to four weeks before the house located on Hyacinth Street was

22 raided on January 20th, 1989. See Exhibit #232, Pgs. 8817, 8818, 8828, 8832. They began residing at

23 the Hyacinth Street house, which was across a field from the said Circle K, about seven months before

24 1/20/89 and moved out three to four months later. See Exhibit #232, Pgs. 8817, 8821, 8822, 8839,

25 Although Jodie Swearingen and Shorty Harden were in and out of the Hyacinth Street house before the

26 raid of 1/20/89, they didn't think that Frank Gable started coming to the Hyacinth Street house until the

27 police raid on 1/20/89. See Exhibit #232, Pgs. 8818, 8819, 8833, 8834. To John Bender's knowledge

28 *Frank Edward Gable vs. State of Oregon Post-Conviction Judgment (a:\PCGable121500.JUD)*   Page -41-

1   he didn't think that Frank Gable and Jodie Swearingen knew each other. See Exhibit #232, Pg. 8818.

2   Kelly Bender did know that Jodie Swearingen had on occasion called Shorty Harden for a ride from the

3   Hyacinth Street house. See Exhibit #232, Pg. 8834.

4   h. Jodie Mae Swearingen disavowed her previous statements to law enforcement officers

5   and others as well as her testimony to the Marion County Grand Jury and testified that Cappie Harden

6   had a reputation for untruthfulness [See Section 8 under this Summary for further details]. Exhibit #232,

7   Pgs. 9324 through 9369.

8   7. Matters relating to the credibility of State's witness Earle Childers:

9   a. Greg Allen Johnson, who was a sales representative of KD Enterprises and on

10   probation, testified that he had a conversation with Earle Childers in January of 1991 when they were

11   in the Marion County Jail. Mr. Childers told Mr. Johnson that he had seen Frank Gable drive off from

12   the Dome Building in what he thought was a Toyota Supra but it could have been a Datsun 280Z. Mr.

13   Childers was sure that the vehicle was brown with black louvers on the rear windows. See Exhibit #232,

14   Pgs. 8624, 8625, 8627.

15   b. Mr. Childers' attorney on new charges in Marion County [Mr. Childers' also had

16   matters pending in Lane County and before the Parole Board], Charles Dean DeMartini, testified about

17   a meeting with the District Attorney's Office and other criminal justice representatives during which

18   pending matters were discussed in the hope of Mr. DeMartini getting some sort of concession for his

19   client, Mr. Childers. Although no concession was obtained, Mr. DeMartini did testify that there was left

20   hanging "some kind of disposition" after the Gable trial was concluded. On cross by the State, Mr.

21   DeMartini did acknowledge that Lane County and the Parole Board had also revoked and imposed time,

22   but Mr. DeMartini said that he did not plan on asking for any sort of further disposition until the prison

23   sentence and Lane County sentence had concluded. See Exhibit #232, Pgs. 8766 through 8773.

24   8. Matters relating to Frank Gable not being at the scene of the murder of Michael Francke:

25   a. Viki Jean Boyd, a friend of Shelli Thomas and Frank Gable, said that she was at

26   Shelli's house at 14th and Vine in Salem when Shelli received a telephone call from Frank Gable

27   between 6:30pm and 7:00pm on January 17th, 1989, and that she talked to Frank Gable on the telephone.

28   *Frank Edward Gable vs. State of Oregon Post-Conviction Judgment (a:\PCGable121500.JUD)*   Page -42-

Yraguen's Decision- pages 43 thru 63

1 See Exhibit #232, Pgs. 8554-8555. She remembers the date and time because she had been at her Parole

2 Officer's office from 4:00pm to 4:45pm, and she remembers that Shelli's boyfriend Dennis Crouse was

3 in jail in California. See Exhibit #232, Pgs. 8556-8557.

4   (1) Shelli Lynn Thomas testified that although she cannot remember the date of

5 the telephone call, she remembers Viki Boyd talking to Frank Gable one of the times he called her house.

6 See Exhibit #232, Pgs. 9397 through 9399.

7   b. Lavonne Joyce Spencer, the Manager of the Chandelle Park Apartments, 3755

8 Hawthorne Avenue NW in Salem, was recalled and again explained that she had served an eviction

9 notice on Mr. Gable the morning of January 18th, 1989, because of "too much noise, traffic in and out,

10 the night before", to wit, January 17th, 1989. Two or three tenants had complained to her the next

11 morning after the noise. See Exhibit #232, Pg. 8655. She then served a second notice of eviction based

12 on the nonpayment of rent. The Gables moved from the Chandelle Apartments on March 3rd, 1989. See

13 Exhibit #232, Pg. 8659.

14   c. Jodie Mae Swearingen, as previously noted, testified for the Defense that she wasn't

15 on the grounds of the Dome Building the evening of January 17th, 1989; that she did not see Frank Gable

16 burglarizing a car that evening; and that she did not see Frank Gable stab Michael Francke. See Exhibit

17 #232, Pg. 9329. She also testified that she did not call Cappie Harden for a ride on January 17th, 1989,

18 and that Cappie Harden, an individual with whom she done drugs and had sex, had a reputation for

19 untruthfulness. See Exhibit #232, Pgs. 9327 through 9329. Ms. Swearingen also testified that she did

20 not meet Frank Gable until the Summer of 1989, although she did acknowledge that she had seen him

21 in 1988. See Exhibit #232, Pg. 9329; 9349.

22   (1) Jodie Mae Swearingen on cross-examination by the State did acknowledge

23 that after she received a written grant of immunity from the State she had many interviews with the

24 police in which she said she did see Frank Gable kill Michael Francke [See Exhibit #232, Pgs. 9334,

25 9336-9337, 9360-9361]; that prior to any police involvement she had told five other persons, to wit,

26 Gary Jensen, Marie Coller, Ann Marie Hagemann, Laura Vorderstraffe, and George Russel Talley, that

27 she was an accomplice and/or had seen the Michael Francke murder [See Exhibit #232, Pgs. 9348

28 *Frank Edward Gable vs. State of Oregon Post-Conviction Judgment (a:\PCGable121500.JUD)*   Page -43-

1 through 9354]; and that she testified before the Marion County Grand Jury that she was present when

2 Michael Francke was killed and that the last time she saw Frank Gable was when Cappie "Shorty"

3 Harden showed up and she got into his car as Frank Gable was running away from the Dome Building

4 [See Exhibit #232, Pg. 9367]. Ms. Swearingen also acknowledged that she had left her Father's place

5 in Dundee on January 17th, 1989, and had placed calls to Cappie Harden on January 17th, 1989, at 11:35

6 a.m., 11:55 a.m., and 12:26 p.m. See Exhibit #232, Pgs. 9339-9340, 9346.

7   9. Matters relating to other individuals being in the vicinity of the Dome Building at the apparent

8 time or at times after the murder of Michael Francke:

9   a. Dianne Lynne Long, who worked in the Department of Corrections personnel records

10 division, testified that she left the Dome Building between 7:02 p.m. and 7:05 p.m. on January 17th,

11 1989, after deciding to leave at 7:00 p.m. according to the clock in her office on the second floor of the

12 Dome Building. See Exhibit #232, Pgs. 9010 through 9014, 9016, 9017, 9024. Her car was located in

13 the North parking lot and as she was leaving the front door of the Dome Building she noticed a silver

14 car, like a Datsun 280Z or something like a sports car, parked directly in front of the Building right up

15 on the sidewalk. See Exhibit #232, Pgs. 9014, 9017. She noted that it was very quiet outside of the

16 Dome Building when she left for the night. See Exhibit #232, Pg. 9022.

17   b. Marsha Ann Haskins was called and her Husband Jim Haskins, was recalled by the

18 Defense. See Exhibit #232, Pgs. 9027 through 9058. Mr. Haskins was head of housekeeping services

19 working out of the basement of the Dome Building, and Mrs. Haskins was at the time employed in

20 housekeeping and the girlfriend of Mr. Haskins. All of the employees had been let go early, to wit, 7:00

21 p.m. rather than 7:30 p.m., and Mr. and Mrs. Haskins were waiting on Wayne Hunsaker so they could

22 leave for a dinner engagement. Mrs. Haskins left the basement of the Dome Building at about 7:05 p.m.

23 to go get her car which was located next to Building 35 via the East basement vehicle service entrance.

24 See Exhibit #232, Pgs. 9035, 9038, 9045. Jim Haskins, as soon as Wayne Hunsaker left via the tunnel

25 door, locked the tunnel door behind Mr. Hunsaker, and then went out the back of the Dome Building

26 via the East basement vehicle service entrance to meet Marsha Haskins. See Exhibit #232, Pgs. 9038,

27 9057. Wayne Hunsaker estimates that he left about 7:10 p.m. See Exhibit #232, Pg. 9052. As Marsha

28 *Frank Edward Gable vs. State of Oregon Post-Conviction Judgment (a:\PCGable121500.JUD)*   Page -44-

1 | Haskins was going to retrieve her car, she noticed a car with four people in it, one of whom got out and

2 | was walking toward the Communications Center which was located in Building 35. See Exhibit #232,

3 | Pg. 9030.

4 | c. Martin Floyd Tooker, a retired police officer, and his Wife Rosella Marie Tooker, were

5 | traveling Westbound on Center Street between 6:50 p.m. and 7:00 p.m. on January 17th, 1989, when

6 | Mrs. Tooker, who was driving their vehicle, nearly struck three (3) individuals who were running in a

7 | Northeasterly direction toward the vicinity of the the Dome Building. See Exhibit #232, Pgs. 9060

8 | through 9062, 9079. Rosella Tooker describes the individual in front as being taller than the other two,

9 | having lighter colored hair, and wearing a tan colored coat to his knees and a light colored hat. See

10 | Exhibit #232, Pgs. 9062, 9063. Marvin Tooker describes the two individuals running behind the light-

11 | haired man as being shorter and dark haired. See Exhibit #232, Pg. 9079. The Tookers received two

12 | (2) threatening telephone calls after they talked to the police telling Mr. Tooker that he'd better keep his

13 | mouth shut and then some calls when no one answered. See Exhibit #232, Pgs. 9089 - 9090.

14 | d. Dale Clayton Harp, who was a passenger in a vehicle being driven by Lisa Zell, his

15 | girlfriend, observed five or six individuals as they were driving by on Center Street between 10:15 p.m.

16 | and 10:30 p.m. running toward a greenish colored Volkswagen van which was parked on 23rd Street.

17 | See Exhibit #232, Pgs. 9462, 9464, 9472. The individuals were about 6 feet tall, had fairly long hair,

18 | were wearing jeans and waist length jeans or leather jackets, and one of the individuals was darker

19 | complexion than the others. See Exhibit #232, Pgs. 9465, 9466.

20 | 10. Other matters relating to the defense:

21 | a. Wayne Scott Hedlund testified that he was employed with Frank Gable at Rigid Truss

22 | from May of 1988 through November of 1988, and that Mr. Gable gave him rides home after picking

23 | up his Wife from the State Hospital. See Exhibit #232, Pgs. 8601, 8603. That when he and Mr. Gable

24 | talked about the Michael Francke murder, Frank Gable didn't know much about it. See Exhibit #232,

25 | Pg. 8603. Mr. Hedlund testified that he saw Mr. Gable with a Swiss Army knife but never with any long

26 | knives. See Exhibit #232, Pg. 8605. In February of 1989, he purchased a trench coat from Mr. Gable

27 | because Mr. Gable needed money for a friend; that Mr. Gable wasn't doing drugs when he knew him;

28 | *Frank Edward Gable vs. State of Oregon Post-Conviction Judgment (a:\PCGable121500.JUD)*   Page -45-

1 and that Mr. Gable had told him that he was a drug dealer but it was only a made-up story. See Exhibit

2 #232, Pgs. 8606, 8610.

3   b. Considerable time, to wit, one complete day of testimony, was spent by the Defense

4 questioning the Oregon State Police Officers, Major Dean Lee Renfrow, Captain Dennis O'Donnell, and

5 Detectives Frederick E. Ackom and John P. McCafferty, in an attempt to establish that there was a

6 purposeful lengthening of the trip from Coos Bay to Salem interrupted by long layovers in order to

7 facilitate questioning of Frank Gable so long as he was willing to continue to talk to the Officers. It was

8 acknowledged that the Oregon State Police Officers did desire to obtain every bit of information possible

9 and during questioning, although Frank Gable never confessed to the murder of Michael Francke, they

10 did feel that there was some slippage, to wit, Freudian slips, on the part of Frank Gable. There was also

11 testimony that Sergeant McCafferty was called to come from Salem to Springfield to continue the

12 interrogation of Frank Gable; the Defense's contention in essence being that there was a desire to

13 pressure Frank Gable into confessing to the crime. See Exhibit #232, Pgs. 8663 through 8747.

14   (1) Richard Kirk Ringler, who had met Frank Gable in late 1988 through Johnny

15 Studer, was called for the purpose of showing that Sergeant McCafferty, when questioning him about

16 the state of his knowledge of matters relating to the Francke murder, "got in his face", and thus by

17 inference may have done the same with respect to his questioning of Frank Gable. See Exhibit #232,

18 Pg. 8755.

19   (2) The Defense wanted to call John Bender and Kelly Bender for the purpose of

20 showing that law enforcement officers subjected them to a "pattern of interrogation designed by the State

21 Police to solve this crime no matter what" [See Exhibit #232, Pg.8794]. After extensive argument

22 before the Court, the Court treated the matter as a motion in limine by the State and granted the motion

23 over objection. See Exhibit #232, Pgs. 8787 through 8798. The desire by the Defense was to show this

24 pattern of interrogation which by inference could be argued as having been the nature of the interrogation

25 of Frank Gable.

26   c. Shelli Lynn Thomas was called by the Defense but, upon advice of her Counsel, she

27 invoked her Constitutional rights not to testify on the basis of self-incrimination. Shelli Lynn Thomas

28 *Frank Edward Gable vs. State of Oregon Post-Conviction Judgment (a:\PCGable121500.JUD)*   Page -46-

1 invoked her rights outside of the presence of the jury. See Exhibit #232, Pg. 8648. Ms. Thomas was

2 later recalled by the Defense and testified that although she could not recall the date, she recalls getting

3 a telephone call from Frank Gable in the early evening when Viki Boyd was at her house at 14th and

4 Hines, and that she spoke with Mr. Gable. See Exhibit #232, Pgs. 9396 through 9398. Ms. Thomas

5 moved back into the 14th and Hines residence as soon as Dennis Gause left for California on January

6 16th, 1989, but she didn't move her furniture back until January 18th when her Mother rented a U-Haul

7 truck. See Exhibit #232, Pg. 9402.

8   d. David Lee Caulley, who was recalled by the Defense, and Mary Blake, both

9 Department of Corrections employees, were questioned in detail by the Defense regarding events

10 associated with Michael Francke and their own actions on January 17th, 2000. See Exhibit #232, Pgs.

11 8877 through 8946. Basically, their testimony coincided with that presented by the State although more

12 details were added as noted under the summary of the State's evidence. In addition, Richard Peterson

13 and Elyse Clawson were also recalled by the Defense.

14   e. In addition to Stanley S. Kudearoff, who was recalled by the Defense, Tyrone Cornelius

15 Williams, Patrick Chas. Boyd, Stephen A. Laknas, and Rocky Lynn Montigue, all of whom passed by

16 the front of the Dome Building at approximately 7:05 p.m. on January 17th, 1989, on their way to the

17 narcotics anonymous meeting at the SOS Club, were called by the Defense and each of the their

18 movements and observations were covered in much greater detail than in the State's case-in-chief. See

19 Exhibit #232, Pgs. 8944 through 9008.

20   (1) Tyrone Williams testified about seeing an individual walking North about fifty

21 (50) yards away. See Exhibit #232, Pg. 8948. Patrick Boyd testified about the individual walking in

22 a strange pattern. See Exhibit #232, Pg. 8959. Stephen Laknas described the individual as being abou

23 5'9", a big man, and wearing dark clothes. See Exhibit #232, Pg. 8973. Rocky Montigue described the

24 individual as being fifty (50) feet away in the grass wearing a dark blue wind breaker, who was abou

25 six (6) feet in height, had short dark hair, weighed about 200 pounds, was stocky and was throwing rocks

26 at a sign, until he glanced at them and started walking away. See Exhibit #232, Pgs. 9002, 9004, 9007.

27   (2) Tyrone Williams saw no other vehicles other than the Francke vehicle an

28 Frank Edward Gable vs. State of Oregon Post-Conviction Judgment (a:\PCGable121500.JUD)   Page -47-

1 saw no vehicle on the West edge of the parking area. See Exhibit #232, Pg. 8947, 8949. Patrick Boyd

2 saw a blue sedan in addition to the Francke vehicle at the scene. See Exhibit #232, Pg. 8961. Stephen

3 Laknas saw no other vehicles in the area other than the Francke vehicle although on cross he

4 acknowledged that another car might have been there. See Exhibit #232, Pgs. 8974, 8976. Stanley

5 Kudearoff reiterated his testimony that he noticed someone on the parking lot who looked like he was

6 looking for his car. See Exhibit #232, Pg. 8989. Rocky Montigue also testified that there were no other

7 cars in the area and maintained that either he or someone else in his group closed the door of the Francke

8 vehicle. See Exhibit #232, Pgs. 9000 through 9002, 9006.

9   f. John D. Middleton, who is employed by the Mental Health Division and works in and

10 was working the evening of January 17th, 1989, in Building #33, which is located next to the Dome

11 Building, heard a louder than usual sound between 6:30 p.m. and 7:00 p.m. which he characterized as

12 like a car accident or glass breaking or a tire iron hitting pavement, to wit, a sharp sound. See Exhibit

13 #232, Pgs. 9215, 9216, 9220. Mr. Middleton looked out his window and saw nothing unusual. See

14 Exhibit #232, Pg. 9217. James Toews, who is also employed by the Mental Health Division in Building

15 #33 and worked until about 9:00 p.m. on budget paperwork and spread sheets, saw people coming and

16 going to the Dome Building from the window in his office but observed nothing and heard nothing

17 unusual during the evening of January 17th, 1989. See Exhibit #232, Pgs. 9223, 9224, 9227.

18   g. At about 7:10 p.m. on January 17th, 1989, Donald Gilbert Moritz had crossed D Street

19 and was walking his German Shepherd dog on the grassy area close to the Dome Building. See Exhibit

20 #232, Pgs. 9283 through 9285. Mr. Moritz observed and heard nothing unusual. See Exhibit #232, Pgs.

21 9286 through 9287. Kerry Dean Thornton was also on the Oregon State Hospital grounds just South of

22 the Dome Building at between 7:15 p.m. and 7:30 p.m. on January 17th, 1989, looking for her Golden

23 Retriever dog and saw and heard nothing unusual. See Exhibit #232, Pgs. 9307 through 9311.

24   h. Richard Carl Clark, who was employed by the Corrections Department and had an

25 office on the second floor of the Dome Building directly above Michael Francke's office, didn't leave

26 work until about 8:00 p.m., and he observed and heard nothing unusual although he noted that the

27 acoustics of the building do not conduct sound very well. See Exhibit #232, Pgs. 9301 through 9305.

28 *Frank Edward Gable vs. State of Oregon Post-Conviction Judgment (a:\PCGable121500.JUD)*   Page -48-

1   Rosilee Ann Breckheimer, who is also a Corrections Department employee and occupies an office on

2   the Northeast corner of the second floor of the Dome Building with a window facing East out the back,

3   did not leave work until about 8:20 p.m. See Exhibit #232, Pgs. 9293, 9297, 9298. Ms. Breckheimer

4   also did not see or hear anything unusual. See Exhibit #232, Pgs. 9293, 9294.

5   i. An attempt was made to call Steve Jackson of the Statesman Journal regarding an

6   article he wrote in February of 1990 resulting from an interview given by Frank Gable in which Frank

7   Gable told Steve Jackson that he said that me and God know I didn't do it rather than the statement taken

8   by Officer Ackom which was termed a Freudian slip. See Exhibit #232, Pg. 9375. However, a motion

9   to quash was filed by the Statesman-Journal and a hearsay objection was interposed by the State. The

10   Court sustained the hearsay objection. See Exhibit #232, Pgs. 9376 through 9382. The Defense then

11   made an offer of proof for the record. See Exhibit #232, Pgs. 9385 through 9387. All of this occurred

12   outside the presence of the jury.

13   j. Sharon Lee Kestermont, who worked in the Construction Division of the Department

14   of Corrections, testified that after the murder, when the employees were allowed to return to the Dome

15   Building on January 19th, 1989, she and three others, Kay Duffey, Karen Trembly and Mary

16   McCullough went out to the North porch, which had supposedly been cleaned, at about 9:00 a.m. and

17   observed blood stains on the porch floor, the left bloody handprint, blood splatters on the wall, blood

18   on the handrail and blood on the door handle, which had not been cleaned up. See Exhibit #232, Pgs.

19   9451, 9452. She stated that she had called the Defense after it appeared that the newspaper reports

20   regarding the blood did not match her recollection. See Exhibit #232, Pg. 9458. Noe Pequeno, who

21   works maintenance in the Dome Building, together with some inmate workers attempted to clean up the

22   North porch. See Exhibit #232, Pg. 9435.

23   k. Bingta Erickson Francke, Michael Francke's widow, was called by the Defense. Mrs.

24   Francke related that at the time of the death of Michael Francke, he was under pressure from the

25   Governor to fire David Caulley regarding some financial issues, and that he didn't want to carry out the

26   Governor's order. See Exhibit #232, Pgs. 9579, 9584, 9585. She also testified that Michael Francke

27   was very security conscious, wore a pager which was not turned off to her knowledge, and had a car

28   *Frank Edward Gable vs. State of Oregon Post-Conviction Judgment (a:\PCGable121500.JUD)*   Page -49-

1 alarm which was working when she left for California one week before his death. See Exhibit #232,

2 Pgs. 9575, 9579.

3   l. Richard H. Fox, a consulting criminalist and from his perspective a quality control

4 expert, was called by the Defense. See Exhibit #232, Pgs. 9608, 9609. Mr. Fox questioned the failure

5 to cordon off the grassy area near the murder scene earlier than was done and the failure to take more

6 blood samples from different areas of the porch including the hand print, hand swipe, and railing. See

7 Exhibit #232, Pgs. 9623, 9632 - 9634, 9636, 9643. He also noted that the physical evidence was not

8 necessarily consistent or inconsistent with the State's theory that a struggle between Michael Francke

9 and his attacker took place near the Michael Francke vehicle, and that there was no physical evidence

10 connecting Frank Gable with the murder scene. See Exhibit #232, Pgs. 9627-9630, 9637.

11   m. Dr. Werner Spitz, a forensic pathologist employed by the Defense, testified that the

12 core body temperature of Michael Francke should have been taken for time of death computations,

13 although he did acknowledge that the results of such temperature readings may widely vary. See Exhibit

14 #232, Pgs. 9665, 9745. He stated that from materials provided in this case, he could not determine the

15 time of death. See Exhibit #232, Pg. 9667. Dr. Spitz testified that none of the knives examined,

16 including the Chicago brand knife, were compatible with the wounds found on Michael Francke's body.

17 See Exhibit #232, Pgs. 9673, 9679, 9681, 9734. His main contention was that the wounds revealed a

18 little cup or half moon immediately under the wound, that is, there was a definite bridge of skin showing

19 a blunt injury as part of the wound, which would indicate that the knife was notched and probably

20 hinged. See Exhibit #232, Pgs. 9674, 9678, 9680, 9683, 9684, 9693, 9724, 9733, 9782, 9783. Dr. Spitz

21 also testified that a 3" to 4" blade could have caused the wounds although he later acknowledged that

22 the blade could have been longer, and that the thickness of the back portion of the knife could have been

23 1/8" although he later acknowledged that he was unable to get a totally accurate measurement of the back

24 portion. See Exhibit #232, Pgs. 9688, 9719, 9721, 9722, 9779, 9734. Dr. Spitz contended that a folding

25 knife or some sort of pocket knife would have been compatible with the wounds inflicted, and that the

26 knife could even have had a serrated blade. See Exhibit #232, Pgs. 9730, 9685, 9696, 9697. Dr. Spitz

27 also spent some time questioning the bruise he said was on the left (right?) forehead of Michael

28 *Frank Edward Gable vs. State of Oregon Post-Conviction Judgment (a:\PCGable121500.JUD)*   Page -50-

1 Francke's body, although he later testified that it was in insignificant blunt force injury (apparently

2 sustained by the body when if fell from the carrier transporting it for autopsy). See Exhibit #232, Pgs.

3 9669, 9671, 9739, 9740.

4

5 *IV. STATE'S REBUTTAL*

6   The State called three (3) witnesses on rebuttal:

7   1. Jill Groves, security manager of US West Communications, produced telephone records

8 relating to Dennis Gause, 1410 Hines St., Northeast, in Salem for January of 1989. See Exhibit #232,

9 Pgs. 9799, 9800, and Exhibit #s 516 and 517. The first telephone call from the Gause residence to

10 Bakersfield, California, to the Kern County Sheriff's main administration office was on 1/21/89 with

11 a second telephone call on 1/21/89 to the Kern County Jail. See Exhibit #232, Pgs. 9803, 9804. On

12 1/23/89 the first telephone collect call received at the Gause residence was at 5:21pm from a presently

13 disconnected number; then a second collect call at 6:03pm from the same presently disconnected

14 number; and then finally a third collect call from 10:39pm from the Santa Clara Jail. See Exhibit #232,

15 Pg. 9805.

16   a. See Vicki Boyd testimony in Defendant's case-in-chief. This evidence was apparently

17 introduced to show that Vicki Boyd and Shelli Thomas would not have even been aware of Dennis

18 Gause being incarcerated in California until on or about 1/21/89.

19   2. Mark D. Ranger, a Detective with the Oregon State Police, testified that he interviewed Shelli

20 Lynn Thomas on 3/6/90, and that she said she first met Frank Gable the day her Mother rented a U-Haul

21 truck. See Exhibit #232, Pgs. 9808.

22   a. See Shelli Thomas testimony in Defendant's case-in-chief. Shelli Thomas testified that

23 her mother rented the U-Haul truck on 1/18/89.

24   3. Richard L. Thorbeck, a Captain with the Department of Corrections, testified that he trained

25 Michael Francke in January of 1988 on the proper use of a shotgun and mini fourteen rifle, because

26 Michael Francke wanted to be trained as a professional. See Exhibit #232, Pgs. 9814, 9820. Michael

27 Francke was subsequently issued an 870 Remington pump shotgun. See Exhibit #232, Pg. 9815.

28 *Frank Edward Gable vs. State of Oregon Post-Conviction Judgment (a:\PCGable121500.JUD)*   Page -51-

1 Captain Thorbeck further noted that Michael Francke expressed no concern about his personal safety

2 in connection with the training or issuance of the shotgun to him. See Exhibit #232, Pg. 9817.

3

4 *V. DEFENDANT'S SURREBUTTAL*

5   No surrebuttal evidence was presented by the Defense. See Exhibit #232, Pg. 9821.

6

7

8

9

10 *VI. SUMMARY OF POST-CONVICTION TRIAL*

11   *FRANK GABLE VS. STATE OF OREGON*

12   *MARION COUNTY CIRCUIT CASE #95-C12041*

13   The post-conviction trial of Frank Edward Gable vs. State of Oregon, Marion County Case No.

14 95C-12041 began on May 1st, 2000, and concluded on May 3rd, 2000, with the Court taking the matter

15 under advisement in order to review the Exhibits, including the 10,500 page transcript of the original

16 trial proceedings as well as for preparation of these written summaries of evidence, findings, conclusions

17 and resulting judgment.

18   In summary, the testimony at the post-conviction trial consisted of the following:

19

20 *Summary of Frank Gable's Case-in-Chief:*

21   1. Matters dealing with Defense Counsel Bob Abel's use of alcoholic beverages prior to trial,

22 during trial, and after trial was concluded:

23   a. Carol I. Wall, a friend of Frank Gable who first met Mr. Gable about a year prior to

24 the beginning of trial and who still visits Mr. Gable on an ongoing basis, testified that during the time

25 when the original trial was in progress, she would on occasion be in the same elevator occupied by Mr.

26 Abel. See Post-Conviction Transcript (abbreviated for further references as "PCT"), Vol. I, Pgs. 45, 48.

27 Carol Wall stated that during those times she could detect a strong smell of alcohol on Mr. Abel. See

28 *Frank Edward Gable vs. State of Oregon Post-Conviction Judgment (a:\PCGable121500.JUD)*   Page -52-

528

1 PCT, Vol. I, Pg. 47. She was unable to observe any other effects upon Mr. Able. See PCT, Vol. I, Pg.

2 47.

3 1) There is also an Affidavit from Freda Amon, a trial observer and former

4 bartender, who swears she saw signs involving Mr. Abel which she felt were characteristic of intoxicated

5 persons, and when she was in the elevator with Mr. Abel he had a "musty smell" that occurs when a

6 person has drank too much and the smell stays in their pores. See Exhibit #13.

7 b. Kevin B. Francke, the Brother of the victim Michael Francke, testified that on one

8 morning during the trial the stench of alcohol was such that it "would knock a maggot off a gut wagon."

9 See PCT, Vol. I, Pg. 54.

10 c. Thos. Nisbet-Lance, a defense investigator, stated that he was in charge of making

11 arrangements for preparation of a model of the Dome Building and when the model was brought to Mr.

12 Abel, he smelled alcohol on Mr. Abel's breath and Mr. Abel was acting "goofy", but acknowledged on

13 cross-examination that Mr. Abel had no other difficulties and he didn't have any personal knowledge

14 about Mr. Abel's reputation for drinking or not recalling details during the late evening hours. See

15 PCT, Vol. I, Pgs. 125, 126, 133.

16 d. Richard Bruce Cummins, a defense investigator, also testified that he smelled the odor

17 of alcohol from Mr. Abel on occasion both before and during trial, and felt that Mr. Abel was under the

18 influence several times during trial although specifics in support of such opinion were not given. See

19 PCT, Vol. I, Pg. 150.

20 e. Roger K. Harris, a defense investigator, testifying about the January, 1991, Depoe Bay

21 meeting for Defense Counsel and Defense Investigators, said that during one evening when a number

22 of the Defense Team, all of whom were drinking, were in the bar, Mr. Abel challenged him to a fight,

23 but then the next morning acted as if nothing had occurred. See PCT, Vol. I, Pgs. 165, 166.

24 f. Cynthia S. Hamilton, a defense investigator, stated that she smelled alcohol on the

25 breath of Robert Abel during pretrial early morning meetings and even obtained private Counsel,

26 William Falls, to discuss what, if anything, she should do about it. See PCT, Vol. II, Pgs. 269, 270. Ms.

27 Hamilton apparently felt that Mr. Abel was impaired although not intoxicated. See PCT, Vol. II, Pgs.

28 *Frank Edward Gable vs. State of Oregon Post-Conviction Judgment (a:\PCGable121500.JUD)*   Page -53-

1 287, 289. Ms. Hamilton also talked to Mr. Storkel about it, and Mr. Storkel acknowledged that Mr. Abel

2 did drink quite a bit but assured her that Mr. Abel always pulled it together for trial. See PCT, Vol. II,

3 Pg. 270. During the one year period she also saw him on two late evening occasions, one of which was

4 at the Depot Bay meeting between 11:00pm and 1:00am, drinking heavily. See PCT, Vol. II, Pgs. 271,

5 272, 286.

6   g. Frank Gable, in his letter to Judge West, dated July 1st, 1991, also notes that "I've

7 smelled alcohol on Mr. Abel's breath during trial." See Exhibit #15. Mr. Gable said that the reason he

8 hadn't brought it up before that time was because of the convict code which involves keeping your

9 mouth shut. See PCT, Vol. II, Pgs. 337, 338.

10   h. Mr. Abel, in his Deposition, states that although he was drinking alcoholic beverages

11 socially, he wasn't drinking immediately before trial proceedings or during trial proceedings and felt that

12 he tried a good case. See Exhibit #3, Pgs. 44, 57, 60. Mr. Abel states that it was after the Gable trial

13 that he started drinking heavily resulting in his entry into a twenty (20) day treatment program in 1994.

14 See Exhibit #3, Pg. 57.

15   (1) John Storkel, Co-Defense Counsel, testified in his Deposition that he did not

16 smell any odor associated with alcoholic beverages on Mr. Abel during trial and did not see any

17 impairment from the consumption of alcohol during trial. See Exhibit #2, Pgs. 37 and 38.

18   2. Matters dealing with the alleged failure to pursue an investigation of Timothy Natividad as

19 a prime suspect in the Michael Francke homicide:

20   a. Kevin B. Francke, after the death of Michael Francke, his Brother, and after his move

21 from Florida to Oregon, married Elizabeth Godlove, who was a former domestic associate of Timothy

22 Natividad before she shot and killed him some time after the death of Michael Francke. See PCT, Vol.

23 I, Pgs. 52, 53, 61. Ms. Godlove, who was represented by Defense Attorney Charlie Burt, was charged

24 and prosecuted for the alleged murder of Timothy Natividad, but was acquitted in a trial which occurred

25 prior to the State vs. Frank Gable trial. Charlie Burt, Elizabeth Godlove nka Francke, and Kevin Francke

26 believed that Timothy Natividad and Scott McAlister, an Assistant Attorney General, who handled legal

27 matters for the Department of Corrections, were somehow responsible for the murder of Michael

28   *Frank Edward Gable vs. State of Oregon Post-Conviction Judgment (a:\PCGable121500.JUD)*   Page -54-

1 Francke. See PCT, Vol. I, Pg. 71. Kevin Francke's beliefs regarding the involvement of Timothy

2 Natividad were apparently based upon the knowledge of his Wife Elizabeth Godlove Francke. Elizabeth

3 Godlove Francke felt that Timothy Natividad was a paranoid drug dealer. See PCT, Vol. I, Pg. 73.

4   (1) Elizabeth Francke's testimony in support of her belief that Timothy Natividad

5 was involved in the murder of Michael Francke, and thus also the basis of Kevin Francke's belief,

6 includes the following:

7   (i) About the time that Michael Francke was killed, Timothy Natividad

8 had a gash on his head and an injury to his leg; See PCT, Vol. I, Pg. 67.

9   (ii) Timothy Natividad had money, carried knives and owned a dark

10 pinstripe suit and a calf-length trench coat (this is obviously a reference to the fact that a man wearing

11 a pin-stripped suit was observed in the Dome Building the afternoon of January 17th, 1989, which turned

12 out to be a Xerox salesman, Dennis Plante, as noted in other portions of the summaries); See PCT, Vol.

13 I, Pgs. 68 and 69. [Note also in these proceedings, that Frank Gable points to a report taken on 8/18/89

14 from Jan Currey, a Department of Corrections employee, contending that the man in the pin-stripped suit

15 was not Mr. Plante – See Exhibit #20]

16   (iii) At some point of time after the death of Michael Francke, Timothy

17 Natividad took his knives to his Brother and told his Brother to clean them up and hold them for him;

18 See PCT, Vol. I, Pg. 69.

19   (iv) She once saw her domestic associate, Timothy Natividad, talking in

20 front of Fred Meyer East in Salem to an unknown male, who, after she was shown a picture of Scott

21 McAlister, she believes was Scott McAlister. See PCT, Vol. I, Pg. 69, 70. She in turn believes that Tim

22 Natividad and Scott McAlister were therefore probably involved in drugs and possibly even child

23 pornography. See PCT, Vol. I, Pg. 73.

24   Also, see Exhibit #214, introduced by the State which is a related 10/3/90 Affidavit of

25 Elizabeth Godlove involving her beliefs regarding Timothy Natividad.

26   (2) Kevin Francke also believes that the State Police conducted a poor murder

27 investigation involving the death of his Brother Michael Francke, and that they should have seized the

28 *Frank Edward Gable vs. State of Oregon Post-Conviction Judgment (a:\PCGable121500.JUD)*   Page -55-

1 Timothy Natividad vehicle, which he says was a 1972 primer gray Chevy Malibu. See PCT, Vol. I, Pgs.

2 55 and 57. Mr. Francke believes that this vehicle was probably the vehicle being operated by Shorty

3 Harden which was seen by Johnny Crouse. See PCT, Vol. I, Pg. 58. Mr. Francke, affirmed by Mrs.

4 Francke, denies that he ever called Robert Abel, as was claimed by Robert Abel in his Deposition for

5 this post-conviction proceeding (See Exhibit #3, Pg. 40), telling him that his Wife Elizabeth would not

6 testify or that she recanted her previous statements to Charlie Burt. See PCT, Vol. I, Pgs. 53, 64, 71.

7   (i) Kevin Francke, on cross-examination, did acknowledge that he had

8 brought a civil action against Frank Gable for the wrongful death of his Brother Michael Francke,

9 because he felt that Mr. Gable was involved in the murder but not the "button man", to wit, the killer.

10 See PCT, Vol. I, Pg. 59.

11   (ii) In essence, Kevin Francke subscribes to a conspiracy theory involving

12 the birth of his Brother Michael Francke, although no evidence has been produced in these post-

13 conviction proceedings which would even remotely suggest the existence of such. As will be noted, the

14 conspiracy theory was considered and investigated by both the State and Defense Investigation Teams.

15   (iii) Mr. Francke also acknowledged that during trial he said that Mr. Abel

16 was turning the State's case into a showing of how not to conduct a murder investigation. See PCT, Vol.

17 I, Pg. 62.

18   b. Thomas McCallum, the lead defense investigator, testified that the Defense

19 Investigators tried to locate the Natividad vehicle but were unsuccessful although they did recover "a

20 bunch of knives and a bunch of other things." See PCT, Vol. I, Pg. 98. Richard B. Cummins, another

21 defense investigator, verified on cross-examination that he viewed the Natividad matter as being a "red

22 herring", although on redirect he said he thought there was lots of other evidence besides the comments

23 of Elizabeth Godlove. See PCT, Vol I, Pg. 157. Mr. Abel mentions the car in his Deposition, but only

24 indicated that Mr. Natividad had several cars and that there had been a long passage of time since the

25 murder of Michael Francke. See Exhibit #3, Pgs. 39 and 42.

26   c. Robert Abel testified in his Deposition that the Defense did receive a tip that Mr.

27 Natividad might have been in a "string" supplying the guards with drugs which were in turn supplied

28 *Frank Edward Gable vs. State of Oregon Post-Conviction Judgment (a:\PCGable121500.JUD)*   Page -56-

1 to the inmates, and that the Defense Team did everything possible to try to track it down. See Exhibit

2 #3, Pg. 38. Mr. Able conversed with Ms. Godlove about Mr. Natividad's involvement in drugs and the

3 murder of Michael Francke, but she had no specifics nor was the Defense Team able to develop any

4 specifics to show any sort of involvement in the murder of Michael Francke. See Exhibit #3, Pgs. 38

5 and 39.

6   d. Although not discussed on the record, the State in it's case-in-chief did introduce

7 materials associated with:

8   (1) The State's investigation of Timothy Natividad [See Exhibit #s 216]. One can

9 also infer from the Defense's summarized discovery results of the investigation and polygraph results

10 of various potential witnesses, as well as many Administrators, Superintendents, and Staff of the

11 Department of Corrections, that the State was attempting to follow up on various conspiracy theories

12 which have swirled around the Gable case. See Exhibit #154.

13   (2) The Defense Team's investigation and pursuit of matters dealing with Timothy

14 Natividad [See Exhibit #s 114, 116, 212, 213, 215, 219, 220, 223, 224, 225, 226, 227, 228, and 229].

15 It is apparent even in these materials that the Defense Team spent considerable time and resources in

16 attempting to pursue the Timothy Natividad matter.

17   (3) The Defense Team's investigation of matters dealing with Jose Navarro [See

18 Exhibit #211].

19   3. Matters dealing with the failure of Frank Gable to testify and/or the desire of Frank Gable to

20 testify during the guilt phase of his trial:

21   a. Paul A. Daholopos, a lawyer and former prosecutor, was hired by the Thomas

22 McCallum to prepare Frank Gable for testifying, in the event that Frank Gable was to be called in his

23 defense, as well as to work on the cross-examination of Jodie Swearingen in the event she was called

24 by the State. Mr. Daholopos' efforts were primarily directed by the investigators, rather than the

25 attorneys, and included a review of materials in order to be able to run Frank Gable through a mock

26 cross-examination Mr. Gable would probably face from the State if called to testify by the Defense. See

27 PCT, Vol. I, Pgs. 39, 42. Mr. Daholopos testified that he believed that such a mock cross-examination

28 *Frank Edward Gable vs. State of Oregon Post-Conviction Judgment (a:\PCGable121500.JUD)*   Page -57-

1   was to be done about May 17th, 1991, as his final preparation for the session occurred on May 16th, 1991,

2   but the lead defense investigator, Mr. McCallum, called telling him that the cross-examination session

3   was postponed. See PCT, Vol. I, Pg. 41. Mr. McCallum testified that it was his understanding that

4   something logistical resulted in the 5/17/91 session being postponed, and that he always though it was

5   going to be rescheduled. See PCT, Vol. I, Pgs. 88, 89.

6   (1) Mr. Daholopos also spent about six hours preparing the Defense for the cross-

7   examination of Ms. Swearingen. See PCT, Vol. I, Pg. 43. However, as noted above in the summary of

8   evidence in the guilt phase of the trial, the Defense, rather than the State, called Ms. Swearingen, and

9   during her testimony for the Defense she basically recanted the testimony she had previously given about

10   being with Cappie Harden when the murder occurred.

11   b. Paul McCallum, the Defense lead investigator, testified that he knew that Frank Gable

12   wanted to testify, and he also knew that Mr. Abel didn't want Frank Gable to testify because he didn't

13   think Mr. Gable would be a good witness. See PCT, Vol. I, Pgs. 87, 88, 89. Thos. Nisbet-Lance, a

14   defense investigator, also felt that Mr. Gable was going to testify. See PCT, Vol. I, Pg. 116. Mr. Nisbet-

15   Lance recalls having a meeting with Mr. Gable, with Tom McCallum present, after the verdict when Mr.

16   Gable told them he had wanted to testify. See PCT, Vol. I, Pg. 122. Richard Cummins, a defense

17   investigator, also testified that he was aware that Mr. Gable wanted to testify, and that his Counsel, Mr.

18   Abel and Mr. Storkel, did not want him to testify. See PCT, Vol. I, Pg. 141. Mr. Cummins' recollection

19   is that a couple of days prior to the Defense resting, there was some sort of confrontation between Mr.

20   Gable and Mr. Abel regarding the subject. See PCT, Vol. I, Pgs. 142, 143. However, on cross-

21   examination Mr. Cummins did acknowledge that he had previously given a statement that Mr. Abel had

22   talked Mr. Gable out of testifying. See PCT, Vol. I, Pgs. 156, 157.

23   c. Cynthia S. Hamilton, a defense investigator, testified that she was aware prior to trial

24   of Mr. Gable's desire to testify and that a decision had been made that Mr. Gable should not testify. See

25   PCT, Vol. II, Pg. 274. Ms. Hamilton wrote a letter to Mr. Abel prior to the beginning of trial that if Mr.

26   Gable was not going to testify, the jury was going to want to know why, and what they would need to

27   do to counter this aspect. See PCT, Vol. II, Pgs. 276, 288 and see Exhibit #234. (Note: This is

28   *Frank Edward Gable vs. State of Oregon Post-Conviction Judgment (a:\PCGable121500.JUD)*   Page -58-

1 numbered 234 but there is also an Exhibit #234 bearing an evidence tag which is the 2/14/90 Oregonian

2 article introduced by the State which is a copy of Ms. Hamilton's undated notes entitled "Thoughts

3 Before Trial") Ms. Hamilton also noted that the Defense Team tried to obtain witnesses who would

4 corroborate some of what Mr. Gable was saying, but that the task had proven difficult and unproductive.

5 See PCT, Vol. II, Pg. 276.

6   d. Frank Gable testified that even though he didn't know his whereabouts on January 17th,

7 1989, at the beginning of trial which would have been in March of 1991, later discussions with

8 investigators led him to know exactly where he was and what had occurred on the January 17th, 1989,

9 and that therefore he should have been called as a witness in his Defense. See PCT, Vol. I, Pgs. 216,

10 217 and see discussion under Findings and Conclusions as what he feels refreshed his recollection and

11 how such information should have been used, now that he has fully reviewed the 10,800 pages of

12 transcript and 29,800 pages of discovery. See PCT, Vol. I, Pgs. 203 through 207, 208, and Mr. Gable's

13 Deposition, Exhibit #156, Pgs. 9 through 21 and 59 through 65. Mr. Gable's position on this particular

14 contention is that this should also have been used in trial as an alibi or explanation of his whereabouts;

15 that is, that even though he didn't know where he was, that the alibi defense should have been interposed

16 because his position was that he was not at the scene of the murder. See Exhibit #156, Pg. 20. Mr.

17 Gable does note in his letter dated July 1st, 1991, to Judge West, which he interprets as a reiteration of

18 his desire to testify and alleged refusal of Defense Counsel to allow such testimony:

19

20 "I requested, and demanded, numerous witnesses to testify on my behalf and was denied this "right" by Mr. Abel also Mr. Abel effectively took away my option to testify on my own behalf by not preparing me, then announcing to the jury that the defense would rest if we could not locate an important

21 witness."

22   (1) Frank Gable testified at another point that he had finally figured out where he

23 was during the evening of January 17th, 1989, about January of 1991, [See PCT, Vol. II, Pg. 358], and

24 that his initial reluctance to testify was because of his prior criminal record being revealed [See PCT, Vol.

25 II, Pg. 330.

26   (2) It should be noted, however, that in Mr. Gable's Deposition introduced by the

27

28 *Frank Edward Gable vs. State of Oregon Post-Conviction Judgment (a:\PCGable121500.JUD)*   Page -59-

1 State, Mr. Gable states that it was late into the trial, two (2) weeks or less of the guilt phase before he

2 finally discovered where he was the evening of January 17th, 1989. See Exhibit #156, Pgs. 18 and 19.

3 Mr. Gable also said in his Deposition that Mr. Able had an adverse reaction to Mr. Gable swearing that

4 he was going to the hospital to deliver a patient drugs. See Exhibit #156, Pg. 26.

5   e. Ms. Karen Steele (Gable), who was Mr. Gable's Federal Defense Counsel and who

6 married Mr. Gable between the guilt and penalty phases of the trial but has subsequently divorced Mr.

7 Gable, was more equivocal in that, although she reiterates that Frank Gable did not initially want to

8 testify (See Exhibit #1, Pg. 9) and she knew Mr. Abel didn't want him to testify, she says in her

9 Deposition that Mr. Gable was waiting to see what witnesses were going to be presented by the Defense

10 and how his attorneys fared with those witnesses (See Exhibit #1, Pg. 13). Mr. Gable subsequently

11 reported to her that he had "not had an opportunity to discuss a final decision with them (Defense

12 Counsel)" before Mr. Able rested the Defense. See Exhibit #1, Pg. 14.

13   f. Robert Abel in his Deposition and at trial simply reiterated his understanding that Mr.

14 Gable "agreed with me that under those circumstances he shouldn't testify." See Exhibit #3, Pg. 49.

15 John Storkel simply noted that Mr. Gable had never told him specifically that he wanted to testify. See

16 Exhibit #2, Pgs. 46 and 47.

17   4. Matters dealing with the Defense Investigators' perception that more time was needed to

18 prepare for trial:

19   a. During January of 1991 the Defense Attorneys, Mr. Abel and Mr. Storkel, and the

20 Defense Investigators, to wit, the Defense Team, held a three-day meeting at Depoe Bay, Oregon, to

21 review the defense of Frank Gable. See PCT, Vol. I, Pgs. 80, 113, 129. Thos. Nisbet-Lance felt that

22 there was a strain between the Defense Attorneys and Defense Investigators which was accentuated by

23 Mr. Abel's romantic involvement with one of the investigators, Jackie Page, and by a potential

24 altercation between Bob Abel and Roger Harris, as noted above. See PCT, Vol. I, Pgs. 113, 114, 135.

25 Richard B. Cummins, a defense investigator, felt that Mr. Abel didn't like him and that he had sort of

26 an antagonistic relationship with Mr. Abel. See PCT, Vol. I, Pg. 139. Roger Keith Harris, a defense

27 investigator, also felt that when he did on occasion have meetings with the Defense Counsel to go over

28 *Frank Edward Gable vs. State of Oregon Post-Conviction Judgment (a:\PCGable121500.JUD)*   Page -60-

1 matters associated with the investigation, Mr. Abel was less approachable than he should have been,

2 and that more time was needed by the investigators to finish investigating the case. See PCT, Vol. I,

3 Pgs. 162, 176, 178, 179.

4 　　　　　(1) About a week after the Depot Bay meeting, Grace Castle, a defense

5 investigator, in conjunction with eight (8) of her colleagues, prepared a letter, a copy of which was not

6 located for purposes of these proceedings, addressed to Judge West which requested a trial

7 postponement. In Mr. Nisbet-Lance's view, the purpose of the letter was to give the Defense Attorneys

8 more time to listen to the Defense Investigators. See PCT, Vol. I, Pgs. 77 through 80, 113, 147. This

9 particular letter was drawn to the Court's attention through formal Motion and Affidavit filed by Mr.

10 Abel and was discussed by the Court. See PCT, Vol. I, Pg. 79; Exhibit #232, Pg. 235. [Also see Exhibit

11 #128 which contains the Motion, Affidavit and Order Denying Motion]. Most of the Defense

12 Investigators felt that the letter upset Mr. Abel, and Mr. Storkel, and that it did lead, in part (it was

13 acknowledged that Ms. Castle's taking vacation on the eve of trial also contributed), to the termination

14 of the employment of Grace Castle. See PCT, Vol. I, Pgs. 76, 77, 115, 148. On the other hand, Ms.

15 Cynthia S. Hamilton, another defense investigator, felt that Mr. Abel had full knowledge of the letter

16 and was in full agreement with it. See PCT, Vol. II, Pg. 269. Ms. Hamilton did sign the letter requesting

17 more time, because she did feel that the Defense Team was dysfunctional, had an unclear focus, and that

18 the Defense Investigators had inadequate access to Mr. Abel which caused stress. See PCT, Vol. II,

19 Pgs. 265, 266, 278.

20 　　　　b. On the other hand, Mr. Able felt that the Defense Team did not need more time; that

21 for a month or so they had hit a dry hole and were not doing anything except to spend money having

22 exhausted every avenue that they had pursued as well as having done everything that the police had

23 done. See Exhibit #3, Pgs. 30 and 31.

24 　　5. Matters dealing Frank Gable's knowledge of the Defense efforts and discovery:

25 　　　　a. Paul McCallum, the Defense lead investigator, had the primary responsibility for pre-

26 trial contact with Frank Gable although other investigators, Richard Bruce Cummins and Roger Harris,

27 would also occasionally meet with Mr. Gable. See PCT, Vol. I, Pg. 83, and see also related Exhibit #119

28 *Frank Edward Gable vs. State of Oregon Post-Conviction Judgment (a:\PCGable121500.JUD)*　　　　　Page -61-

1   which was an Order Allowing Investigator Visits during any weekday or weekend up to 10:30pm except

2   for standard meal times. Paul McCallum had up to ten (10) investigators working on the defense (Mr.

3   Able in his Deposition says that there were thirteen (13) investigators at one time -- See Exhibit #3, Pg.

4   11), all of whom reported to him. See PCT, Vol. I, Pg. 76. There were about 30,000 pages which Mr.

5   McCallum reviewed which involved about six months of work.

6   (1) Mr. McCallum recalls that Frank Gable would mainly be briefed on the

7   defense investigation efforts and only occasionally be asked to read a particular report(s). See PCT, Vol.

8   I, Pg. 84. Prior to trial Mr. McCallum had weekly meetings with the Defense Attorneys. See PCT, Vol.

9   I, Pg. 82. Mr. McCallum felt that Mr. Able was opinionated and was particular about wanting things

10   done his way. See PCT, Vol. I, Pg. 82. The Defense Attorneys mainly met with Frank Gable in the

11   morning during trial before trial would begin. See PCT, Vol. I, Pg. 87. Mr. McCallum would meet with

12   Mr. Gable after trial in the evening. See PCT, Vol. I, Pg. 112. Mr. Gable in his Deposition stated that

13   Mr. Storkel visited him two to three times with "these big book things", and that he read quite a bit of

14   discovery during the jury selection process although he is a slow reader. See Exhibit #156, Pgs. 23 and

15   24.

16   (2) Richard B. Cummins recalls meeting with Mr. Gable hundreds of times and

17   going over discovery, police reports and defense reports with him, although the reports were not actually

18   left with him. See PCT, Vol. I, Pgs. 139, 140.

19   b. John Storkel, Co-Counsel for Mr. Gable, in his Deposition swears: 1) That Frank Gable

20   made the determination that Jodie Swearingen would be called as a witness over the objection of Mr.

21   Abel (See Exhibit #2, Pg. 31); 2) That throughout trial Mr. Gable had the bottom line authority as to

22   whether a witness would or would not be called (See Exhibit #2, Pg. 31); 3) That Mr. Gable had

23   continuing input in both the guilt and penalty phases of trial (See Exhibit #2, Pg. 19); and 4) That Mr.

24   Gable told Mr. Abel, to his chagrin, that Mr. Storkel was going to take over as Counsel during the

25   penalty phase which took place as per Mr. Gable's directions (See Exhibit #2, Pgs. 13 through 15).

26   6. Matters relating to Frank Gable's potential alibi or whereabouts defense:

27   a. Paul McCallum felt that Frank Gable was pretty much able to tell them his whereabouts

28   *Frank Edward Gable vs. State of Oregon Post-Conviction Judgment (a:\PCGable121500.JUD)*   Page -62-

1   on January 17th, 1989, and early morning hours of January 18th, 1989, even though he had told police he

2   didn't know for sure where he was and couldn't specifically recall to them where he was that night. See

3   PCT, Vol. I, Pg. 90, 91. Mr. McCallum also felt that they had some credible evidence of Mr. Gable's

4   whereabouts and that there was some discussion with Mr. Abel about an alibi notice. See PCT, Vol. I,

5   Pg. 92. However, Mr. McCallum recognized that there were holes in an alibi defense. See PCT, Vol.

6   I, Pg. 102. Thos. Nisbet-Lance did locate a parking ticket for a time close to January 17th, 1989, to wit,

7   January 16th, 1989, which might have helped in reviving Mr. Gable's memory. See PCT, Vol. I, Pg. 128.

8   b. Mr. Gable feels that Defense Counsel failed to discuss strategies with him and failed

9   to allow him to see all of the discovery as opposed to that discovery that they "gleaned down" for his

10   defense. See PCT, Vols. I & II, Pgs. 212, 213, 243. See also as noted above, under the section dealing

11   with the Defense's case-in-chief in these post-conviction proceedings, and more specifically under Mr.

12   Gable's desire to testify, the tie between the failure to testify Claim and the alibi or whereabouts Claim.

13   Mr. Gable also points to his letters to Judge West, as notification to the Court of his disagreements with

14   how he was being and had been represented. See Exhibit #s 14 and 15. The first letter, Exhibit #14, was

15   written on March 31st, 1991, Mr. Gable expresses concern that the defense had not been given adequate

16   time to prepare the defense, and that Mr. Gable was not getting all of the information about his defense,

17   which was the basis for his requested continuance. It should be noted, however, that this occurred

18   midway through the voir dire portion of trial, and after Mr. Gable was reviewing the discovery "gleaned

19   down" for his review [see note also to this effect in Exhibit #15], as well as after hearing the questioning

20   by the State and his Defense Counsel of potential jurors. The second letter, Exhibit #15, was written on

21   July 1st, 1991, the day that the penalty phase was initially set to begin before being continued by the

22   Court. The July 1st, 1991, letter expresses Mr. Gable's disgust at the incompetence shown by his

23   Defense Counsel and not being prepared to testify, or as Mr. Gable interprets his words, being allowed

24   to testify by the Defense.

25   (1) Mr. Gable did address the Court about the 7/1/91 letter stating: "Yeah. I have

26   one more matter for the record. I would like to state for the record that at this point I'm really not happy

27   that we have reached this point. I would like to refer the Court back to my letter that I sent to you

28   *Frank Edward Gable vs. State of Oregon Post-Conviction Judgment (a:\PCGable121500.JUD)*   Page -63-

Yraguen's Decision- pages 64 thru 84

1 previously. I feel it's those issues in that letter, why we're at this point today." The Court replied:

2 "Okay. Thank you. We'll note that statement for the record." See Exhibit #232, Pg. 10082. See also

3 proceedings regarding the letter on 7/2/92 – Exhibit #232, Pg. 10085.

4   7. Other matters relating to Post-Conviction Relief issues:

5   Premature and/or Improper Discovery to the State

6   a. Paul McCallum stated that discovery to the State was primarily handled by Grace

7 Castle, after such discovery had first been reviewed by the Defense Attorneys. See PCT, Vol I, Pg. 95.

8 The only report that he can recall the Defense turning over to the State that he didn't think should have

9 been released was a report of one of the investigators following Kevin Francke around for a couple of

10 days, but in Mr. McCallum's view that did no damage to the Defense's case. See PCT, Vol. I, Pg. 103.

11 Mr. McCallum thought that there might have been other things which were erroneously turned over to

12 the Defense but couldn't recall any specific items. See PCT, Vol. I, Pg. 105. On the other hand, Robert

13 Abel in his Deposition swore that the reports regarding the surveillance on Kevin Francke were not

14 turned over to the State. See Exhibit #3, Pg. 34. Richard B. Cummins, another defense investigator,

15 also testified that some discovery regarding Jodie Swearingen was given to the State prematurely in his

16 view since the Defense didn't have to give up the information until they determined that they were going

17 to put her on the stand. See PCT, Vol. I, Pg. 152. Mr. Cummins' point appeared to be that because the

18 State had the defense discovery, they were better able to conduct cross-examination of Ms. Swearingen

19 when she was called by the Defense regarding statements she had made in Denver, which statements

20 were not followed up on with the Denver witnesses. See PCT, Vol. I, Pg. 153.

21   Improper Handling of Jodie Swearingen as a Witness

22   b. Paul McCallum recalled that Mr. Abel did not want Mr. Storkel to do redirect

23 examination on Jodie Swearingen, because Mr. Able felt she was too dangerous and too unpredictable

24 a witness. See PCT, Vol. I, Pg. 96. Jodie Swearingen's Father was also not called regarding Jodie,

25 Swearingen being in Dundee until the late evening hours of January 17th, 1989, because he was not

26 viewed as being a very credible witness. See PCT, Vol. I, Pg. 97. Mr. Cummins, a defense investigator,

27 who actually interviewed Jodie Swearingen's Father can't recall having an opinion about his credibility

28 *Frank Edward Gable vs. State of Oregon Post-Conviction Judgment (a:\PCGable121500.JUD)*   Page -64-

1   as a potential witness, although his only purpose would seem to have been for the impeachment of Jodie

2   Swearingen's past statements implicating Frank Gable which would then undermine Cappie Harden's

3   testimony from Mr. Gable's perspective. See PCT, Vol. I, Pg. 144. Mr. Cummins' opinion was that

4   Mr. Storkel had no concept of the facts of the case, and although Mr. Abel had a better understanding

5   of the facts, the Defense Counsel split up the witnesses and thus didn't know as much as they should

6   have known about each other's witnesses. See PCT, Vol. I, Pg. 145. Also, Mr. Cummins seemed to

7   imply a complaint about Mark McKnight, the Attorney for Jodie Swearingen, writing up the questions

8   to be asked by Mr. Storkel on direct examination of Jodie Swearingen by the Defense, and, in turn, the

9   fact that there was no redirect examination of Jodie Swearingen. Mr. Cummins also complained about

10   Mr. Storkel at the Depoe Bay meeting referring to a newspaper accounts of what witnesses to be called

11   by the Defense were going to say. See Pct, Vol. I, Pgs. 145, 146.

12   (1) It was noted in Exhibit #176 submitted by the State that Jim Swearingen,

13   Jodie's Father, told Detective Mark Ranger on 1/30/90 that he did not recall if Jodie Swearingen was

14   at his residence when he arrived home the afternoon of 1/17/89; and Jerry Yarborough, apparently Mr.

15   Swearingen's domestic associate, believed that she had lunch with Mr. Swearingen and that Jodie

16   Swearingen was gone when she arrived home. On 7/20/90 they both declined to speak with the Defense.

17   See Exhibit #175.

18   Grand Jury Participation by Thomas Denney, an Assistant Attorney General

19   c. Dale W. Penn, the Marion County District Attorney at the time of the Michael Francke

20   prosecution testified regarding the Grand Jury composition which returned the Indictment against Frank

21   Gable. Mr. Penn noted that even though Thomas Denney of the Appellate Division of the Department

22   of Justice had in past years given periodic updates on appellate cases to the District Attorneys, Mr.

23   Denney was not associated with the investigation of the Michael Francke case other than as a member

24   of the Grand Jury. See PCT, Vol. I, Pgs. 190 through 196. The District Attorney's Office had nothing

25   to do with the selection of the individual Grand Jurors which process was under Court auspices and

26   control, and further, that there was no process for objecting to a Grand Jury member for the reasons

27   being put forth in this case. See PCT, Vol. I, Pgs. 192, 193.

28   *Frank Edward Gable vs. State of Oregon Post-Conviction Judgment (a:\PCGable121500.JUD)*   Page -65-

1

<u>Summary of Frank Gable's Major Issues</u>

2   d. Frank Gable's testimony revolved around his approach to what he regarded as primary

3 issues which were interrelated and included, but not limited to:

4   (1) An explanation of his whereabouts or alibi for the night of January 17th, 1989,

5 referred to as the "Gable timeline";

6   (a) The "Gable timeline" (See Exhibit #9) includes an attack on the

7 credibility of Jodie Swearingen (even though Jodie Swearingen was called by and testified for the

8 Defense that she had lied and knew nothing and saw nothing the night of Michael Francke's murder)

9 which in turn would involve an indirect attack on the credibility of Cappie Harden. These recollections

10 would then have placed Mr. Gable in a position to testify which he says he wanted to do from the

11 beginning of trial but was hampered by his continuing claim that he did not know his whereabouts the

12 night of January 17th, 1989.

13   (2) Impeachment of Janyne Gable, his former Wife, that Frank Gable was not at

14 home and was gone the entire night of January 17th, 1989.

15   (3) Impeachment of Cappie Harden by bringing in evidence that until after the

16 time of the raid on the Bender house which occurred on January 20th, 1989, Cappie Harden didn't know

17 Frank Gable. This evidence would be in Mr. Gable's view in the form of witnesses like Ronald Ruis,

18 Janet Sanderson, and Sam Harmon.

19   (4) Other miscellaneous matters.

20   In summary, Frank Gable testified as follows:

21   (1) Whereabouts or alibi for night of January 17th, 1989:

22   Frank Gable contends that if the Oregon State Police Officers who were

23 questioning him on 11/3/89 had told him that the date of the eviction notice, to wit, 1/18/89, that would

24 have caused him to recollect the basis for the eviction which was unnecessary noise involving a plate

25 throwing incident attributed to Scott Dowd which could have been verified by Mark Gessner. See PCT,

26 Vol II, Pgs. 224 through 226, 234, 235.

27   The "Gable Time Line", in this Court's best attempt to give some sort of order

28 *Frank Edward Gable vs. State of Oregon Post-Conviction Judgment (a:\PCGable121500.JUD)*   Page -66-

1   to it from Mr. Gable's Notebook, his testimony at the post-conviction relief trial, and his Deposition,

2   begins with Frank Gable contention that if he had been shown the parking ticket, which he told Defense

3   Investigator Bruce Cummins about, that he received on 1/16/89 (See Exhibit #4) at the Salem Memorial

4   Hospital, that would have triggered his recollection that he made three (3) trips to the Hospital to visit

5   Mike Weaver, a prison inmate, the first being with Kenny Farrell and Janyne Gable, the second being

6   alone when the got the ticket, and the third being on 1/17/89 when he learned that Mike Weaver had

7   been returned to the prison hospital when he saw Gordon Martin and Wayne Knepper in the parking lot.

8   See PCT, Vol. I, Pgs. 231, 232. Mr. Gable says in his Deposition that he was taking heroin shots to

9   Mike Weaver which were being supplied by Kenny Farrell. See Exhibit #156, Pgs. 10 through 13.

10   This in turn coupled with Vicki Boyd's and Shelli Thomas' testimony would

11   have caused him to recollect that he first talked to Vickie and Shelli by telephone on January 17th, 1989;

12   then went to the Hospital finding that Mike Weaver was gone; then went to visit Vicki and Shelli during

13   which time he made a drug deal with Mark Gesner for Robert Cornett and that Chris Warilla and Mickey

14   Goss later arrived at that same residence. Frank Gable accompanied by Mickey Goss then meet Robert

15   Cornett and Lamont Lee at an arcade; then Frank Gable and Mickey Goss go back to the Gable house

16   where he saw various individuals Kevin Walker and Mark Gesner. See PCT, Vol. I, Pg. 218.

17   (a) The attack on Jodie Swearingen is a bit more convoluted in that, in

18   spite of how Jodie Swearingen testified for the Defense, it could have been shown from Frank Gable's

19   view, that at the short visit between Jodie Swearingen and Cappie Harden, they got their stories to

20   coincide. Then by the Defense getting into evidence the 3/12/90 police report in which Jodie

21   Swearingen told police that she used a stolen credit card to place the telephone calls to Cappie Harden

22   from the Plaid Pantry, which is a fourteen (14) minute walk from the Dome Building and thus from the

23   phone records and time sequence could not have occurred as she said it did, he would have discredited

24   her past statements to law enforcement authorities. See PCT, Vol. II, Pgs. 245, 252, 254. In essence

25   taking the same tact as the cross-examination in getting into her prior statements, but then attack the

26   particulars of those statements. This also would have involved calling Jodie Swearingen's Father as well

27   as attacking her statements involving her being at Dundee on 1/17/89, and the times involved in making

28   *Frank Edward Gable vs. State of Oregon Post-Conviction Judgment (a:\PCGable121500.JUD)*   Page -67-

1 the trip back to Salem. See PCT, Vol. II, Pgs. 296, 297, 300. While this would have constituted an

2 attack on her credibility, the attack would then carry over indirectly to the testimony of Cappie Harden.

3   (2) Impeachment of Janyne Gable:

4   (a) Frank Gable complains that although the Defense reserved their

5 questioning of Janyne Gable, it never followed up on such. Mr. Gable feels that Michael Goss should

6 have been used to testify that he spent the whole night at Mr. Gable's with respect to a report generated

7 by Defense Investigator Cynthia Hamilton. See PCT, Vol. II, Pg. 303. In addition, Scott and Anise

8 Dowd should have been called to testify regarding the plate throwing incident which Mr. Gable contends

9 occurred the night of 1/17/89. See PCT, Vol. II, Pgs. 303, 304. Mr. Gable also believes that even

10 though Lavonne Spencer, his landlady, was called, her testimony should have been handled in a more

11 effective manner. In addition, Mr. Gable believes that Chris Warilla could have testified that the big

12 drug deal, 1/4 pound of speed, took place on 1/17/89 and that others such as Lamant Lee, Mickey Goss

13 and Robert Cornett could have also testified to events occurring on 1/17/89. See PCT, Vol. II, Pgs. 308,

14 311. Mr. Gable also believes the phone calls associated with his house at 2:33pm, 4:22pm and 5:38 pm

15 were associated with Kevin Walker calling about a drug deal. See PCT, Vol. II, Pg. 310. Finally, with

16 respect to knives being taken from the Gable household, he believes that Armando Garza was the one

17 taking knives and that this could have been established to discredit that portion of Janyne Gable's

18 testimony. See PCT, Vol. II, Pg. 318.

19   (3) Impeachment of Cappie Harden:

20   (a) Frank Gable testified that the Defense Counsel should have called

21 witnesses, other than the nine (9) witnesses who were called, to attack Cappie Harden's credibility. Mr.

22 Gable supplied names, including Rebecca Dimbat, Larry Pilgrim, Ronald Ruis, Janet Sanderson, Verna

23 Jo Hensley, Leann Battern, Chris Warilla, Sam Harmon and the Confidential Reliable Informant used

24 by the police to secure a search warrant for the Bender house as being potential witnesses for the

25 Defense. See PCT, Vol. II, Pgs. 245 through 251, 257, 292, 293, 295. All of the named individuals

26 were acknowledged by Mr. Gable to be part of the drug scene [See PCT, Vol. II, Pg. 368], but Frank

27 Gable feels that even though the Benders testified that they did not feel that Mr. Gable was at the Bender

28 *Frank Edward Gable vs. State of Oregon Post-Conviction Judgment (a:\PCGable121500.JUD)*   Page -68-

1   house until after the police raid of 1/20/89 and even though they did not think that Jodie Swearingen and

2   Frank Gable knew each other, these other individuals could have testified similarly and the Defense was

3   ineffective by not calling these additional witnesses.

4   (4) Other miscellaneous matters:

5   (a) Frank Gable was married to Karen Steele, his Federal Defense

6   Attorney, on July 10th, 1991. See PCT, Vol. I, Pgs. 201, 215. Frank Gable contends that, contrary to

7   Bob Abel's deposition, Karen Steele's visits were limited when he was in jail in Marion County and that

8   Defense Counsels' access to him was not affected by his romantic involvement with Karen Steele. See

9   PCT, Vol. I, Pg. 215, 349. On the other hand, Robert Abel in his Deposition was of the opinion that

10   Karen Steele became lead counsel in penalty phase of the trial, because Mr. Gable listened to her and

11   no one else at that stage (See Exhibit #3, Pgs. 9 and 10), and John Storkel in his Deposition felt that

12   Karen Steele "created an unknown factor in dealing with the client that you would prefer not to have in

13   a defense case (See Exhibit #2, Pg. 13).

14   (b) Frank Gable has prepared materials involving the alleged involvement

15   of Timothy Natividad. See Exhibit #8. However, in his testimony Frank Gable complained about Bob

16   Abel's pursuance of "his little conspiracy theory trip" [See PCT, Vol. II, Pg. 236], which the Natividad

17   matter clearly appears to be; that is, Frank Gable appeared to be implying that time was being wasted

18   by his Defense team. Frank Gable also contends that Timothy Natividad was the pin-suited man seen

19   at the Dome Building on 1/17/89. See PCT, Vol. II, Pgs. 319 through 321. A review of materials

20   contained in Mr. Gable's Exhibit #8 did not reveal any new information other than was known at the

21   beginning of trial regarding the alleged involvement of Timothy Natividad in the murder of Michael

22   Francke. Mr. Gable in his Deposition also complained about Mr. Abel's "multiple assailant theory" and

23   wanting to become a super star by proving a conspiracy. See Exhibit #156, Pgs. 4, 35 and 36.

24   (c) Frank Gable also prepared materials showing Johnny Crouse's

25   involvement in the Michael Francke murder. See PCT, Vol. II, Pgs. 322, 323 and Exhibit #7. A review

26   of materials contained in Mr. Gable's Exhibit #7 also did not reveal any new information other than was

27   known at the beginning of trial regarding the alleged involvement of Johnny Crouse in the murder of

28   *Frank Edward Gable vs. State of Oregon Post-Conviction Judgment (a:\PCGable121500.JUD)*   Page -69-

1 Michael Francke.

2   (d) Frank Gable also contended that other individuals who were in the

3 vicinity of the Dome Building, Richard D. Carey, who works at Ward 40 and saw nothing, and Kenny

4 Thornton, who was walking his dog, should have been called. See PCT, Vol. II, Pg. 260. Contrary to

5 Mr. Gable's position, however, Kenny Thornton was called by the Defense.

6

7 *Summary of State's Case-in-Chief*

8   1. Matters relating to Robert Able's use of intoxicating beverages:

9   a. Sarah Moore Bostwick and Thomas C. Bostwick both testified that even through they

10 were at counsel table right next to Mr. Abel, with Ms. Moore being right next to Mr. Abel, neither felt

11 that Mr. Abel showed any sign of intoxication nor did she smell any odor associated with alcoholic

12 beverages on Mr. Abel. See PCT, Vol. II, Pgs. 401, 410, 417, 418. In fact, Mr. Bostwick tried two other

13 murder cases against Mr. Abel in 1989 and 1990, and he never noticed any signs of drinking during

14 either of those trials either. See PCT, Vol. II, Pg. 419.

15   b. The presiding Circuit Judge, Greg West, testified that a rumor, which could have even

16 been before trial, that Mr. Abel was drinking did come to his attention through his judicial assistant,

17 Diane Boehmer, which resulted in Judge West calling in Mr. Able and telling him about the rumor. See

18 PCT, Vol. III, Pgs. 550, 551. Judge West did not observe Mr. Abel to be impaired by alcohol and had

19 no concern about such during trial. Judge West also noted that if anyone has any question about Mr.

20 Abel's state, there is a film of the entire trial. See PCT, Vol. III, Pg. 552, 558, 559. Ms. Boehmer

21 cannot recall how the matter was brought to her attention. See PCT, Vol. III, Pg. 568. Diane L.

22 Boehmer, a judicial assistant who was present in the courtroom during the Gable trial, did not detect the

23 odor of alcohol on Mr. Abel and did not feel that he was impaired at any time during trial. See PCT,

24 Vol. III, Pg. 565.

25   c. Robert Abel testified that he was not drinking heavily and was not impaired by his

26 drinking during the Frank Gable trial. See PCT, Vol. III, Pg. 542. Mr. Able did acknowledge that he

27 began drinking heavily after the Gable trial and did attend a twenty (20) day alcohol treatment program

28 *Frank Edward Gable vs. State of Oregon Post-Conviction Judgment (a:\PCGable121500.JUD)*   Page -70-

1   in 1994. See PCT, Vol. III, Pg. 490.

2   2. Matters relating to Grand Jury proceedings:

3   a. Sarah Moore Bostwick and Thomas C. Bostwick both acknowledged working with

4   some members of the Department of Justice including Brenda Peterson of the Appellate Division and

5   Randy Martinek, Katherine McLaughlin, Chuck Pritchard and Bob Hamilton of the District Attorneys

6   Assistance Division, but both testified that Tom Denney had nothing to do with the case other than being

7   a member of the Grand Jury. See PCT, Vol. II, Pgs. 406, 420. Sarah Moore testified that the Grand Jury

8   was actually requested in August of 1989 to assist in the investigation and justify the huge expenditure

9   of investigative resources being committed by the Oregon State Police. See PCT, Vol. II, Pg. 390

10   through 392. Ms. Moore felt that they had no basis in fact or law to ask for the removal of Tom Denney

11   from the Grand Jury. See PCT, Vol II, Pgs. 392, 393,

12   b. Sarah Moore acknowledged that Detective Pierce was allowed to be present at Grand

13   Jury proceedings pursuant to a request to the Court which resulted in a Court Order allowing Detective

14   Pierce to attend the Grand Jury proceedings. See PCT, Vol. II, Pg. 390 and Exhibit #101..

15   3. Matters relating disclosure of plea agreements with witnesses:

16   a. Sarah Moore Bostwick and Thomas C. Bostwick both testified that no plea agreements

17   or deals were involved with anyone which weren't disclosed to the Defense. See PCT, Vol. II, Pgs. 394,

18   413, 414.

19   b. Robert Able acknowledged that to his knowledge the State never failed to disclose any

20   plea agreement that they had with a witness. See PCT, Vol. III, Pg. 489. The only complaint of John

21   Storkel was not that there was any nondisclosure of any plea agreement but that some witnesses were

22   treated "awfully good by the State". See Exhibit #2, Pgs. 34 and 35.

23   c. Exhibit #157 shows with specificity materials provided to the Defense which delineate

24   those witnesses with whom there were considerations made for their cooperation with the investigation

25   of the homicide of Michael Francke as well as the criminal histories of witnesses the State intended to

26   call.

27   4. Matters relating to true life consideration:

28   *Frank Edward Gable vs. State of Oregon Post-Conviction Judgment (a:\PCGable121500.JUD)*   Page -71-

1   a. Sara Moore Bostwick and Thomas C. Bostwick both acknowledged that discussion of

2 the true life option was raised with the Trial Court. See PCT, Vol. II, Pgs. 408, 422. See Exhibit #232,

3 Pgs. 2796 through 2799, 3000 through 3004.

4   (1) Mr. Bostwick stated on the record in part as follows: "*** I have discussed

5 it with counsel for the defense is that this particular instance was committed prior to the passage of that

6 new section which allows the jury to impose true life without the possibility of parole (See Exhibit #232,

7 Pg. 2796, Lines 14 through 18) *** I don't want Mr. Storkel addressing the jury about the three options

8 in the penalty phase and then having the defense at a later date raise the issue and successfully argue to

9 the Court that true life is not an option in this particular case and then have the jury misinformed at this

10 stage (See Exhibit 232, Pgs. 2797, Lines 6 through 11)." Mr. Storkel replied "We'll discuss it with our

11 client.. I also told Mr. Bostwick at this point I didn't think that it was going to be an issue, so I would

12 like to continue to be able to question the jury about that. Mr. Gable has been charged under the new

13 statute and so, therefore, I think that the three options are the way that the Court would probably rule.

14 But I'll talk with Mr. – Mr. Gable and if he decides that way then there will be no issue of controversy."

15 See Exhibit 232, Pg. 2797, Line 19, through Pg. 2798, Line 2. The Court concluded that the State

16 "***has agreed that you can discuss it with them (jury)." See Exhibit #232, Pg. 2799, Line 14. Mr.

17 Gable was present at the April 1st, 1991, proceedings when the noted discussion took place.

18   (2) When the matter again came up on April 2nd, 1991, Mr. Abel said in part:

19 "Your Honor, we have not discussed the matter with our client as of this time. However, my position

20 is going to be, and I have talked to Mr. Storkel about this, we don't want to make a decision at this stage

21 of the proceedings and, therefore, neither one of us will be addressing that issue with prospective jurors

22 further." See Exhibit #232, Pg. 3001, Lines 4 through 9. There then appeared to be general concurrence

23 that it was not an issue until the penalty phase. See Exhibit #232, Pg. 3004, Lines 9 through 13. Mr.

24 Gable was present at the April 2nd, 1991, proceedings when the noted discussion took place.

25   b. Robert Abel in his testimony simply recalled the question being raised. See PCT, Vol.

26 III, Pg. 533.

27   c. Presiding Circuit Judge Greg West did recall the matter being raised and concluded

28 *Frank Edward Gable vs. State of Oregon Post-Conviction Judgment (a:\PCGable121500.JUD)*   Page -72-

1 the choice was made to go with life without parole as a defense strategy to give jurors a viable choice

2 other than the death penalty. See PCT, Vol. III, Pg. 548, 555 through 557. Judge West also provided

3 copies of his opening instructions preceding the penalty the afternoon before the penalty phase was to

4 begin and advised Counsel that he would plan on using those instructions unless he heard otherwise from

5 Counsel. See Exhibit #232, Pg. 10082. The same penalty provisions were given by the Court at the

6 beginning of the penalty phase as were given in the final instructions quoted below. See Exhibit #232,

7 Pgs. 10087 and 10088.

8   d. John Storkel in his Deposition initially stated that he did not know until after trial that

9 there was an ex post facto objection involving the instruction of life without parole when he presented

10 the three options (death; life without parole; life with a 30 year minimum) for jury consideration. See

11 Exhibit #2, Pgs. 41 and 42. This error in recollection was later corrected by Mr. Storkel in an Affidavit

12 dated 4/31/2000. See Exhibit #231. Mr. Storkel does not add any further information regarding the

13 status of the issue in the said Affidavit. Mr. Abel on April 2nd, 1991, tells the Court he has discussed

14 the matter with Mr. Storkel. John Storkel in his Deposition further states that his perception was that

15 the Jury might have gone with the death penalty if they only had two choices, to wit, life with a 30 year

16 minimum or death. See Exhibit #2, Pg. 42.

17   (1) It appears also that Mr. Storkel submitted proposed Jury Instructions dated

18 July 10th, 1991, which were filed on July 16th, 1991, which included proposed Instruction No. 8 as

19 follows: "Oregon law provides that the penalty of Aggravated Murder shall be either life imprisonment

20 or death. In this case, I instruct you that you must consider that life imprisonment means imprisonment

21 for the rest of the defendant's life without possibility of probation, parole, work release or early release."

22 See Exhibit #151.

23   e. The Court during its final penalty phase instructions instructed in part as follows:

24 "If you answer one or more of the questions no, the law requires that the penalty shall be life
imprisonment without the possibility of parole or release unless ten or more members of the jury find

25 there are sufficient mitigating circumstances to warrant life imprisonment with the possibility of parole
or release.

26 "A person sentenced to life imprisonment without the possibility of parole or release shall not
have that sentence suspended, deferred or commuted by any judicial officer. And the State Board of

27 Parole may not parole the prisoner or reduce the period of confinement in any manner whatsoever.

28 *Frank Edward Gable vs. State of Oregon Post-Conviction Judgment (a:\PCGable121500.JUD)*   Page -73-

1 "The State Department of Corrections or any executive official may not allow the prisoner to participate in any sort of release or furlough program.

2 "Life imprisonment with the possibility of parole or release means that the Court must order the defendant to be confined for a minimum of thirty years without the possibility of parole, release or work

3 release or any form of temporary leave or employment at a forest or work camp. After twenty years in prison, the defendant would be able to petition the State Board of Parole for an order altering his

4 sentence to life imprisonment with the possibility of parole or work release. The defendant would have the burden of convincing all five members of the Parole Board that he would be likely to be rehabilitated

5 within a reasonable period of time before his sentence could be altered." See Exhibit #232, Pgs. 10512 and 10513.

6

       f. The Jury upon the returned verdict voted "no" on question four eliminating the penalty

7 of death. The Jury voted "no" on question five which asked if there were sufficient mitigating

8 circumstances to warrant life imprisonment with the possibility of parole. See Exhibit #232, Pgs. 10516

9 and 10522. Mr. Gable waived time and asked to be sentenced immediately (See Exhibit #232, Pgs.

10 10524, 10525), and the Court imposed the sentence of life without the possibility of parole (See Exhibit

11 #232, Pg. 10526).

12

13   5. Matters relating to Frank Gable testifying:

14   a. Mr. Gable was vacillating about testifying, and Robert Abel testified that he talked

15 Frank Gable out of testifying before the Defense rested. See PCT, Vol. III, Pgs. 477, 478, 501 through

16 503. Robert Abel told Frank Gable that it would be stupid for him to testify (See PCT, Vol. III, Pg. 537)

17 because:

18   (1) Mr. Abel didn't believe, and still doesn't believe, that the State had proven

19 their case (See PCT, Vol. III, Pg. 475);

20   (2) The Defense had worked on trying to come up with some sort of alibi defense

21 for Mr. Gable as from Mr. Gable's possible scenarios and their investigations, including names (Chris

22 Warilla, Kevin Walker, Mark Gessner, Robert Cornett, Mickey Goss) given to them by Mr. Gable, were

23 producing nothing of substance (See PCT, Vol. III, Pgs. 471, 472, 473, 474, 482, 483, 484, 541);

24   (a) In addition, producing more drug dealers and addicts simply

25 emphasized the "jockey boxing" theory of the State as to what Mr. Gable was doing in Mr. Francke's

26 vehicle in the first instance (See PCT, Vol. III, Pgs. 480, 481);

27

28 *Frank Edward Gable vs. State of Oregon Post-Conviction Judgment (a:\PCGable121500.JUD)*   Page -74-

1   (b) Mr. Abel testified that "we never, ever were able to put something

2 together that we could say he was here and he did this, this and this"; in essence, the Defense was never

3 able to close the window of opportunity to commit the murder (See PCT, Vol. III, Pgs. 483, 484)

4   (3) Mr. Abel was concerned about what Mr. Gable's demeanor would be on the

5 stand (See PCT, Vol. III, Pg. 476);

6   (4) Mr. Gable was never able to get his story together (See PCT, Vol. III, Pg. 498);

7 and

8   (5) Finally, a question arose as to what advantage, if any, would be gained by

9 showing that the plate incident and noise for the eviction was attributed to others since the landlady had

10 been called and testified that it was because of the Gables' noise, which would show that Janyne Gable

11 wasn't at home alone as she said (See PCT, Vol. III, Pg. 521).

12

13   6. Matters relating to Natividad/Crouse investigations:

14   a. Robert Abel testified that he refused Kevin Francke's request for access to the Gable

15 Defense files; that Kevin Francke had called and told him that Elizabeth would not testify; and that in

16 any event he did not feel that Elizabeth Godlove was a credible witness. See PCT, Vol. III, Pgs. 465,

17 515. The Timothy Natividad matter was investigated and produced no substantive information. See

18 PCT, Vol. III, Pgs. 463, 467, 469, 470 as well as Exhibits noted in above summary of the Gable case-in-

19 chief.

20   b. Robert Abel testified that the Johnny Crouse matter was pursued without any success

21 in being able to get the matter into evidence during the trial due to adverse Court rulings. See PCT, Vol.

22 III, Pgs. 485, 486, and Exhibit #s 209 and 210.

23

24   6. Matters relating to Gable Defense investigation, preparation and discovery:

25   a. Robert Abel testified that he reviewed roughly 30,000 pages of the State's discovery,

26 and went through the initial discovery material three (3) different times. See PCT, Vol III, Pg. 430. The

27 Defense made two complete sets of the discovery they received from the State -- one intact set exactly

28 *Frank Edward Gable vs. State of Oregon Post-Conviction Judgment (a:\PCGable121500.JUD)*   Page -75-

1 as they received it and a working copy. See PCT, Vol. III, Pg. 432. Both he and Mr. Storkel set up

2 computer retrieval systems -- his for selected materials and Mr. Storkel for everything received. See

3 PCT, Vol III, Pgs. 431, 432, 439. See also a mass of materials introduced by the State which include

4 Defense Notebooks, Summary Notes, Time Lines, Legal Research, etc., in the form of Exhibit #s 152,

5 153, 155, 194 through 205, 207, and 230 (multi-volumned). The materials alone show a huge

6 expenditure of time and resources by the Gable Defense Team and Defense Counsel.

7   (1) Monday morning meetings were set up with the investigators and assignments

8 were given to them by Mr. Abel and updates received from them, and then the investigators then had

9 their own separate follow up meeting. See PCT, Vol. III, Pg. 435 and See Exhibit #159 showing

10 examples of Defense Team Agendas and matters discussed. Tom McCallum was the lead investigator

11 and would take most of the significant reports to Mr. Gable for review although no materials were left

12 with Mr. Gable. See PCT, Vol. III, Pg. 459. Mr. Able met with Mr. Gable most mornings prior to trial.

13 See PCT, Vol. III, Pg. 459.

14   (a) Mr. Abel did feel that Karen Steele was a distraction and a bother for

15 the Defense and for Mr. Gable. See PCT, Vol. III, Pgs. 462, 512, 513.

16   (2) A law clerk, criminalist, pathologist, psychologist, psychiatrist, forensic

17 documents examiner, forensic graphics expert and investigators (up to 13 investigators at one point)

18 were hired and utilized by the Defense. See PCT, Vol. III, Pgs. 441, 442 and related Exhibit #s 108, 109,

19 111, 115, 117, 118, 120, 121, 122, 129, 131 through 149 .

20   (3) Robert Abel acknowledged that he used a trial tactic of having investigators

21 contact potential witnesses but not producing any written report because he, in particular, did not want

22 to memorialize any negative information to the Gable defense. See PCT, Vol. III, Pgs. 436, 437. The

23 Defense turned discovery over to the State but waited on some of it until just before trial began. See

24 PCT, Vol. III, Pg. 445.

25   (a) This information seems to be corroborated by Grace Castle's Affidavit

26 (See Exhibit #6), although it needs to be recognized that this Affidavit was prepared after she was fired

27 and as a general attack on Defense Counsel. Ms. Castle swears that the Defense Investigators were

28 *Frank Edward Gable vs. State of Oregon Post-Conviction Judgment (a:\PCGable121500.JUD)*   Page -76-

1   instructed to write reports in a manner which would include nothing which would be harmful to Mr.

2   Gable. See Exhibit #6, Pg. 8. Ms. Castle went so far as to write the investigative reports and then to

3   separately write confidential memorandums to Defense Counsel. See Exhibit #6, Pg. 8. Ms. Castle was

4   upset when she learned as part of the discovery process to the State, one of her confidential

5   memorandums was disclosed to the State. See Exhibit #6, Pg. 8. Apparently the Depoe Bay meeting,

6   which produced the letter to Judge West, also involved serious disagreements between Defense Counsel

7   and the Defense Investigators about what had to be disclosed to the State with the Defense Counsel

8   taking a more expansive approach. See Exhibit #6, Pgs. 8 and 9.

9   b. Mr. Able testified that although various investigators had their pet theories, those

10 theories were producing no results. See PCT, Vol. III, Pg. 508. The Defense did follow up on the

11 State's dead ends, but also just came up with a bunch of garbage. See PCT, Vol. III, Pg. 446. Mr.

12 Storkel in his Deposition simply concluded that the Defense Investigators were trying to make decisions

13 and determinations that were decisions and determinations to be made by Defense Counsel. See Exhibit

14 #2, Pg. 25.

15   (1) Ms. Castle's Affidavit was also more revealing than was the testimony of the

16 individual Defense Investigators during the post-conviction trial, in that she swears that the main

17 difficulty between Defense Counsel and the Defense Investigators arose after the Depoe Bay meeting

18 because of a chronology prepared by Tom Lance in which he outlined the activities of nearly 200

19 persons, gleaned from police tip sheets and his personal notes, which were names not given to the

20 Defense in the State discovery, which he and the Defense Investigators felt should be pursued. Defense

21 Counsel did not give serious consideration to the Lance materials. See Exhibit #6, Pg. 6.

22

23   7. Matters relating to Jodie Swearingen:

24   a. Robert Abel testified that he did not want to call Jodie Swearingen, but Frank Gable

25 insisted that she be called. See PCT, Vol. III, Pg. 488. Once Jodie Swearingen recanted and said

26 everything she had said was a lie, Mr. Abel felt that nothing was going to be gained by having Jodie

27 Swearingen acknowledge that she made additional statements tying Frank Gable to the murder of

28 *Frank Edward Gable vs. State of Oregon Post-Conviction Judgment (a:\PCGable121500.JUD)*   ·   Page -77-

1 Michael Francke, and finally, the Defense was not in a position to turn around and attack the veracity

2 of their own witness when they wanted the Jury to accept the testimony which she had given on direct

3 examination. See PCT, Vol. III, Pgs. 488, 489.

4

5   8. Matters relating to Janyne Gable:

6   a. Robert Abel testified that after Janyne Gable testified and the Defense reserved the

7 right to call her at a later time, he reinterviewed Janyne Gable and decided against recalling her because

8 it was apparent that she was going to remain firm in her testimony. See PCT, Vol. III, Pgs. 522, 539.

9 Mr. Able also felt that even though they did have phone records of calls associated with the Gable house,

10 those phone records did not establish the presence of Mr. Gable. See PCT, Vol. III, Pg. 523.

11

12   9. Other miscellaneous matters:

13   a. Sara Moore Bostwick testified, as it relates back to the Natividad matter, that the man

14 in the pin-stripped suit seen in the Dome Building the afternoon of January 17th, 1989, turned out to be

15 a Xerox repairman. See PCT, Vol. II, Pgs. 399, 400. Mr. Abel also acknowledged that their

16 investigation revealed that the man wearing the pin-stripped suit turned out to be Dennis Plant or Pliant,

17 a Xerox salesman. See PCT, Vol. III, Pgs. 468, 517.

18   b. Robert Abel did acknowledge that he ordered Grace Castle, a defense investigator, to

19 be fired after she took a vacation on the eve of trial. See PCT, Vol. III, Pg. 447.

20   c. Robert Abel did testify that the letter from the defense investigators to Judge West did

21 upset him; that he disagreed with it; and that the continuing investigation was producing little in the way

22 of results, but that he agreed to submit it to Judge West and did so by Motion and Affidavit for

23 continuance. See PCT, Vol. III, Pgs. 448 through 451.

24   d. Near the conclusion of the State's case-in-chief the State offered Exhibit #s 102

25 through 234, which were received without objection. See Exhibit #232, Pg. 567. The said Exhibits

26 contain law enforcement reports, defense investigator reports, Court filings, Defense notebooks, and a

27 mass of other materials. Where applicable or pertinent in this Court's view, some of those Exhibits have

28 *Frank Edward Gable vs. State of Oregon Post-Conviction Judgment (a:\PCGable121500.JUD)*   Page -78-

1   been noted in this summary of the post-conviction relief proceedings as well as under the specific

2   Findings and Conclusions set forth below.

3

4   *Summary of Frank Gable's Rebuttal*

5   1. Frank Gable resumed the stand and testified that he wrote a letter to Judge Lipscomb wanting

6   a continuance because he wanted additional investigation to be done including interviews with Scott and

7   Anise Dowd, Gordon Martin, Kenny Farrell, Lamont Lee, Robert Cornett, Sam Harmon and Wayne

8   Knepper, but Mr. Abel was ready to go to trial. See PCT, Vol III, Pgs. 570 through 573, 578, 579. Mr.

9   Gable doesn't know whether these individuals would have helped him or not, but his point appears to

10   be that the Defense Team did not do enough to learn what they might have to add to the proceedings and

11   therefore he was ineffectively defended. Frank Gable also wanted to explain his only God and I know

12   who killed Michael Francke statement, and testified that he had been up endless hours with no sleep

13   when the statement was made. Mr. Gable felt that Mr. Abel was doing a "shit-ass job" with it and so

14   he refused to come to Court one day. See PCT, Vol. III, Pgs. 574 through 577. Frank Gable again

15   reiterated the inconsistencies of the statements of Jodie Swearingen, and the Defense's failure to follow

16   up on those statements. See PCT, Vol. III, Pgs. 581 through 584. Basically, Mr. Gable's rebuttal was

17   a reiteration of matters raised in his case-in-chief.

18   2. Randy C. Martinak, a Department of Justice investigator, called Mr. Hadley and wanted to

19   testify because he had evidence which would impeach Mr. Abel. See PCT, Vol. III, Pg. 585. When Mr.

20   Martinak was allowed by this Court to testify by telephone, his testimony actually had to do with his

21   part in the Johnny Crouse investigation, and how he felt that the wrong man, Mr. Gable, was being

22   accused. See PCT, Vol. III, Pgs. 596, 597. It appeared that Mr. Martinek felt that the Defense Attorneys

23   should have asked him his opinion about the case when they attempted to get into the Crouse matter.

24   In any event, nothing of substance was added to the record.

25

26   **VII. POST-CONVICTION CLAIMS IN FRANK GABLE'S THIRD AMENDED COMPLAINT**

27

28   *Frank Edward Gable vs. State of Oregon Post-Conviction Judgment (a:\PCGable121500.JUD)*   Page -79-

1 | ***SPECIFIC FINDINGS AND CONCLUSIONS OF FACT AND LAW ON SEPARATE CLAIMS***

2

3 | ***FIRST CLAIM OF RELIEF***

4 | *CLAIM #1:* IT IS ALLEGED THAT DEFENSE COUNSEL ROBERT L. ABLE AND JOHN E. STORKEL FAILED TO MEET WITH AND ADVISE DEFENDANT GABLE ABOUT THE

5 | DEFENSE'S INVESTIGATION.

6 | *FINDINGS and CONCLUSIONS:*

7 | a. Defendant Gable appeared with Defense Counsel at pre-trial hearings, many of which dealt with discovery matters, on fifteen (15) occasions, to wit, 4/12/90, 5/10/90, 7/3/90, 7/10/90, 7/11/90,

8 | 8/28/90, 9/6/90, 9/12/90, 10/3/90, 11/5/90, 1/11/91, 1/23/91, 2/4/91, 2/5/91, and 2/21/91.

9 | (1) It is noted that Defendant Gable at the February 4th, 1991, hearing, told that Trial Judge that he didn't even need to be present at the next hearing [See Exhibit #232, Pg. 245], but the Trial

10 | Judge was of the opinion that he needed to attend Court when matters of "substance" were discussed. This seems to be a strange position for a Defendant who now says he was so totally in the dark as to

11 | what was being done on his case.

12 | b. It also appears that Defense Counsel were listening to Mr. Gable's desires. For example, when during a brief break in trial, Defense Counsel advised the Court that Mr. Gable wanted to proceed with

13 | trial without delay against the advice of counsel. See Exhibit #232, Pg. 7634. John Storkel's Deposition further illustrates the point through some highlighted areas of the Defense which were controlled by Mr.

14 | Gable which included: 1) Frank Gable's determination that Jodie Swearingen would be called even though Defense Counsel were skeptical of calling her; 2) Frank Gable's bottom line authority as to

15 | whether a witness would or would not be called; and 3) Frank Gable's determination that Mr. Abel would be replaced by Mr. Storkel during the penalty phase of the trial. See Exhibit #2, Pgs. 13 through

16 | 15, 31. Mr. Storkel also noted that Mr. Gable had input in both the guilt and penalty phases of trial. See Exhibit #2, Pg. 19.

17

18 | (1) Mr. Abel acknowledges, with evident disgust in his Deposition, that Mr. Gable, in conjunction with Karen Steele in Mr. Abel's view, had significant control of the communication and even with his removal for all intensive purposes from the penalty phase of the trial. See Exhibit #3, Pgs.

19 | 25 and 46.

20 | (2) The record does reflect that Mr. Storkel did indeed take over the penalty phase contrary to his previous expectations. Even during that penalty phase we find the following dialogue

21 | taking place between the Court and Mr. Gable, when Mr. Storkel put on the record that Defense Counsel were advising Mr. Gable that he should not testify and Mr. Gable had decided to follow their advice (See

22 | Exhibit #232, Pg. 10310):
    Mr. Gable: "Also, I had an issue that I have talked with Mr. Storkel about and I believe,

23 | it's my understanding, that Mr. Abel was going to do the closing, and I have requested that Mr. Storkel do it. And I would like that put on the record."

24 | The Court: "It's your request that Mr. Storkel do the closing remarks?"
    Mr. Gable: "Yes, it is."

25 | Mr. Able: "And I was aware of that and had no intention of going against Mr. Gable's wishes. I will not do the closing." See Exhibit 232, Pg. 10311, Lines 10 through 20.

26

27 | (3) Mr. Gable does, when Defense Counsel are putting on the record their disagreement with putting Karen Steele on the stand during the penalty phase, which Mr. Gable was demanding, state

28 | *Frank Edward Gable vs. State of Oregon Post-Conviction Judgment (a:\PCGable121500.JUD)*   Page -80-

1 that "in the first stage of the trial, there were many witnesses I did want called and weren't called" to explain his demands in the penalty phase. See Exhibit #232, Pg. 10314 (see 10312 through 10314 for
2 entire discussion regarding Karen Steele).

3   c. In spite of even this last statement from Mr. Gable, it is difficult to understand or accept how someone exercising the control over matters which Mr. Gable was exerting could be unaware of the
4 pertinent parts of the investigation and how the Defense was going to be handled and was being handled.

5   d. Mr. Gable acknowledged during his testimony in the post-conviction trial that he was getting discovery when he testified he was concerned about the defense because "staff***investigators were
6 telling me and some of the reports they were letting me read." See PCT, Vol. II, Pg. 331 and PCT, Vols.
   I and II, Pgs. 212, 213, 243. Mr. Gable in his Deposition notes that Mr. Storkel visited him two or three
7 times with these big book things (obviously materials being organized for trial purposes), and that he read quite a bit of discovery during selection of the jury which took over six (6) weeks. See Exhibit
8 #156, Pgs. 23 and 24.

9   e. Mr. Gable was present, as noted, during all of the pretrial hearings as well as trial proceedings including opening statements. It is a reasonable inference that Mr. Gable was not only aware of what
10 the Defense was doing but also what the Defense planned to do in defending him against the murder charges.
11   f. Mr. Gable's position appears to simply be that he did not get all of the 30,000 pages of discovery – as opposed to only some of the discovery that the Defense planned on using. See PCT, Vol.
12 II, Pg. 368. It wasn't necessary or feasible to get all 30,000 pages of discovery to Mr. Gable nor has Mr.
   Gable been able to raise much of substance even now after having had ample opportunity to go through
13 all 30,000 pages of discovery.

14   g. In addition, all of these findings and conclusions directly relate to the issue of whether Mr. Gable was denied an opportunity to testify during the guilt phase of the trial, as that claim will be
15 discussed in more detail below.

16

17 *CLAIM #2:* IT IS ALLEGED THAT DEFENSE COUNSEL ROBERT L. ABEL AND JOHN E. STORKEL FAILED TO MEET WITH DEFENDANT GABLE BEFORE AND DURING TRIAL TO
18 PLAN AN EFFECTIVE DEFENSE STRATEGY.

19 *FINDINGS and CONCLUSIONS:*

20   a. The failure to be effective appears to be based upon the result, to wit, the fact that Mr. Gable was convicted of the charges which were brought against him, rather than some sort of inappropriate
21 defense strategy. The defense strategy appears to have included: 1) testing every aspect of the State's case; 2) showing through numerous witnesses and exhibits why the State's witnesses should not be
22 believed; 3) showing the presence and/or existence of other possible suspects or individuals who had the opportunity to commit the crime; and 4) stressing the State's inability to connect Mr. Gable with
23 the crime scene other than through witnesses who should not be believed by the Jury; to wit, the absence of any trace evidence connecting Mr. Gable with the crime scene.

24
25   b. The findings and conclusions set forth under Claim #1 above are also applicable under this Claim. Mr. Gable was fully aware of what the Defense was doing and had considerably more influence and control than he is now contending.

26
27   c. The most difficult challenge for the Defense was how to deal with statements made by Mr. Gable to so many different individuals. In other words, one of the greatest hurdles the Defense faced

28 *Frank Edward Gable vs. State of Oregon Post-Conviction Judgment (a:\PCGable121500.JUD)*   Page -81-

552

was Mr. Gable's mouth. Mr. Gable not only made incriminating statements to many different individuals under many different circumstances, he even attempted to play a "cat and mouse" type game with the Oregon State Police, and at one point even invited the police to meet with him regarding their investigation. It was almost as if Mr. Gable was saying to the police – if you tip enough of your hand to me, I will be able to outsmart you and/or verbally destroy whatever you think you have on me. Mr. Gable was his own worst enemy in building the State's case against himself, but even at this stage of proceedings, Mr. Gable still fails to recognize this obvious deduction when one reviews his case in any sort of detailed analysis such as is being done in these post-conviction relief proceedings.

(1) Mr. Gable did testify in these post-conviction proceedings. In large part, the testimony was an attempt to verbalize his way past the evidence presented against him at trial, in large part ignoring any incriminating evidence as being lies and/or including an attempt to discredit certain named individuals, some of whom testified and some of whom did not testify, and then constructing an "alibi" for himself, some of the "alibi" which is linked to some portions of the evidence given during trial but much of the "alibi" simply being Mr. Gable talking without much surmising as to who would say what to support him without showing any specificity of the assertions he made. And just as Mr. Gable did with law enforcement officers and various associates before trial, Mr. Gable speaks with assurance that, just as his mouth allowed him to handle anything he encountered before the murder of Michael Francke, his mouth will set him free in these proceedings. What Mr. Gable has instead accomplished is that he has added yet another version of events on January 17th, 1989, which although more elaborate than many of his past versions is still full of "holes".

CLAIM #3: IT IS ALLEGED THAT DEFENSE COUNSEL ROBERT L. ABEL AND JOHN E. STORKEL FAILED TO GIVE AN ALIBI NOTICE EVEN THOUGH DEFENDANT GABLE HAD ALWAYS DENIED BEING AT THE SCENE OF THE DEATH OF MICHAEL FRANCKE.

FINDINGS and CONCLUSIONS:

a. Frank Gable gave many different versions of his whereabouts on January 17th, 1989, including an ongoing inability to recount his whereabouts, as noted in the statements taken by law enforcement officers as noted in the above Summaries. The statements taken from Mr. Gable by his Defense Attorneys also involved the same inability to recount his whereabouts with any sort of specificity. See Exhibit #208 which has transcribed interviews between Mr. Abel and Mr. Gable on 8/23/90, 9/13/90, 11/2/90 and one dated 1/17/89 which would appear to have an erroneous date on it since the appointment of Mr. Abel did not occur until 4/9/90. It should also be noted from the Summaries that even though the Defense Investigators were attempting to corroborate Mr. Gable's versions of events right up to the conclusion of the guilt phase, even they were forced to acknowledge his latest version " holes" still existed.

b. Mr. Gable's contention during the post-conviction trial was that although he didn't know his whereabouts at the beginning of trial, as the trial developed he finally realized where he was. This is particularly interesting, since as early as 2/14/90 in a newspaper interview published by the Oregonian, Mr. Gable was advising the news media that he had three (3) witnesses who could prove exactly where he was on 1/17/89 at the time of the murder. See PCT, Vol. II, Pg. 361, and Exhibit #234. Yet, even in that same newspaper article, it was noted that Frank Gable had one week earlier given a statement to the Salem Statesman Journal in which he said "Does anybody remember what they were doing on a specific night a year ago? Add to that the fact that I was in the bag, strung out on crank, and had been up several days, 24 hours a day, with some friends. I don't even know which day was which."

(1) Consider even in matters involving these post-conviction relief proceedings, that Mr. Gable in his Deposition taken on 12/18/98 says that it was late into trial, to wit, two weeks or less of the guilt phase remaining before he finally realized where he was on January 17th, 1989. See Exhibit #156,

1 Pgs. 18 and 19.

2   c. As noted, Mr. Gable in this post-conviction relief trial contends that he now has it all together in what he has termed the "Gable Time Line". See Exhibit #9. His present recollection is now

3 apparently based upon two events, to wit, 1) a parking warning which he received at 1:30pm on January 16th, 1989, at the Salem Hospital [See Exhibit #4, Pg. 21 of 49], and 2) an eviction notice received on

4 January 18th, 1989, from the Gables' landlady, Lavonne J. Spencer.

5   (1) Ms. Spencer testified in both the State's and the Defense's cases during trial about the eviction notice and did indeed testify that the eviction was due to too much noise the night before,

6 to wit, the night of January 17th, 1989.

7   (2) Mr. Gable looks to Scott and Anise Dowd as being the cause of the noise due to the two of them fighting and a plate being shattered, and that he was actually in the middle of a dope deal

8 upstairs with Chris Warilla when the argument and plate busting occurred. Scott and Anise Dowd were not called during trial nor were there any Affidavits or testimony in this post-conviction trial of what

9 their testimony would be if called. In a January 2nd, 1991, telephone call from Alaska, it is apparent that Scott Dowd's implied concurrence that the fight with his Wife took place on 1/17/89 comes from the

10 Defense Investigator Cummins saying that Mr. Gable is saying that that is when the fight and plate incident took place. Mr. Dowd, who did not even learn of the eviction notice until sometime later,

11 simply takes that date as given to him and then discusses the details. Ms. Spencer simply related that there was a lot of traffic in and out and noise of people running up and down stairs, which Ms. Spencer

12 may indeed have mistakenly attributed to the Gables, since she was in bed when the noise occurred and did not get up to determine who specifically was responsible for it. (See Summaries of Lavonne Spencer

13 testimony at trial; Exhibit #9, and, in particular, Discovery Pages 028296, 002415 through 00249 and Pg. 11 of 10/22/90 interview apparently by a Defense Investigator). In essence, the date of the Dowds'

14 argument is far from settled.

15   (3) Mr. Gable also is apparently contending that the big drug deal on the night of 1/17/89 was with Chris Warilla. This is the same Chris Warilla who, as noted in the Summaries above, was

16 involved in a continuous relationship of buying and selling drugs with Frank Gable since June or July of 1988.

17

18   (i) Although this Court understood that Chris Warilla was supposedly the individual with whom Mr. Gable had the big drug deal the night of 1/17/89, in going through the

19 Exhibits, this Court made note of comments emphasized in Exhibit #16, Pg. 2, a report by Ms. Castle, a Defense Investigator, in which it appears that Mr. Gable is viewing Mr. Gesner as being the individual

20 with whom he had the big drug deal, although he may indeed be saying that that deal was at the Shelli Thomas' residence instead of his own residence. The report does have the following: "When I asked

21 him if he remembered a plate being thrown through the window, he said "That happened after I left." Nowhere in the report are any comments attributed to Mr. Gesner which says that any of the occurrences

22 were the night of 1/17/89 – all that appears to come from Mr. Gable's now refreshed recollection.

23   (ii) It appears that Chris Warilla, Mark Gesner, and Frank Gable were so heavily involved in drugs and drug dealing that it would be unlikely that any one of these individuals could with

24 specificity delineate the times and dates of their many transactions.

25   (iii) Chris Warilla did not testify at trial, but if he had been called as contemplated by Mr. Gable, it is probable that one more individual would be speaking about Mr. Gable telling him

26 that he had stabbed a man in a parking lot when he was in the man's car. See Exhibit #s 182 and 183. The Defense Team apparently had difficulty in their attempts to locate Mr. Warilla. See Exhibit #184.

27   (4) Mr. Gable also looks to the testimony of Vicki Boyd and Shelli Thomas, both of

28 *Frank Edward Gable vs. State of Oregon Post-Conviction Judgment (a:\PCGable121500.JUD)*   Page -83-

1 whom were called by the Defense about the telephone call between them and Frank Gable allegedly occurring on 1/17/89, but in any event, after Dennis Gause was arrested in California. However, from
2 a time perspective, their testimony was placed in considerable question in the State's rebuttal case, and it appears likely that any such telephone call occurred well after 1/17/89 particularly since Mr. Gause
3 was not even arrested in California until 1/19/89. See Exhibit #232, Pg. 8856. See also both State and Defense investigative reports #s 160, 161, and 164 through 171, which appear to support this Court's
4 conclusion that the testimony of Vicki Boyd and Shelli Thomas is subject to considerable doubt.

5   (5) Mark Gesner and John Kevin Walker are also part of the "Gable Time Line". This is the same Mark Gesner who related nothing about being at Frank Gable's apartment on 1/17/89 and
6 during his testimony at trial said that Frank Gable had told him in July of 1989 that he had stuck Michael Francke three or four times; and the same John Kevin Walker who related during testimony at trial that
7 he was indeed in Salem in the early morning hours of 1/18/89 but not at Frank Gable's apartment and that Frank Gable had told him the next day during the delivery of drugs to him by Mr. Walker that he
8 had stuck Michael Francke and if Mr. Walker ever told he would kill him.

9   (6) Other names have been given by Mr. Gable such as Robert Cornett, Lamont Lee, and Mickey Goss, Kenny Farrell but again we have no Affidavits or testimony regarding Mr. Gable's
10 contentions in his recollected memory, and these names are the same given to the Defense by Mr. Able who were checked out and produced nothing of substance. See PCT, Vol. III, Pgs. 471 - 474, 482-484,
11 541. Materials involving Mickey Goss are discussed in more detail under Mr. Gable's attack on the credibility of his former Wife, Janye Gable.

12
13   (i) The Defense reports in the evidence regarding Kenny Farrell do not support Mr. Gable's contention of his whereabouts on January 17th, 1989. See Exhibit #s 186 and 187.

14   d. The point of all this is, however, not to establish the precise facts of the movements and actions described, because that has been a task attempted by both the State and Defense without great
15 success, but rather that: 1) Evidence was given during trial regarding what is now called the "Gable Time Line"; and 2) The only new aspect during these post-conviction proceedings is that this is simply newest
16 rendition, admittedly with more details, from Mr. Gable of what occurred the night of 1/17/89. Although Mr. Gable would like to categorize his Exhibit #9 as an "alibi defense", it is does not rise to
17 that sort of legal defense. It was much as Mr. Abel concluded when asked on Deposition with respect to not raising an alibi defense, "there was nothing credible" which warranted raising such a defense, and
18 "we never, ever were able to put something together that we could say he was here and he did this, this and this." See Exhibit #3, Pg. 28, and PCT, Vol. III, Pgs. 483, 484. That same status exists today as it
19 did in 1991 when Mr. Gable's case was being tried.

20   e. Finally, Mr. Gable's Deposition of 12/18/98 for these post-conviction relief proceedings is revealing in that Mr. Gable takes the position that his Defense Counsel were required to file an alibi
21 defense for him when he took the position that he wasn't at the scene of the crime even if he couldn't remember where he was at the time of the crime. See Exhibit #156, Pg. 20. Mr. Gable does not seem
22 to be aware of the requirements of ORS 135.455 regarding the notice required for an alibi defense which includes a requirement that the alibi defense "shall state specifically the place or places where the
23 defendant claims to have been at the time or times of the alleged offense together with the name and residence or business address of each witness upon whom the defendant intends to rely for alibi
24 evidence." It simply was not practical or possible for Defense Counsel to proceed in the manner now suggested by Mr. Gable.

25

26 CLAIM #4: IT IS ALLEGED THAT DEFENSE COUNSEL ROBERT L. ABEL AND JOHN E. STORKEL FAILED TO READ THE REPORTS AND DISCOVERY FURNISHED BY THE STATE
27 OF OREGON.

28 *Frank Edward Gable vs. State of Oregon Post-Conviction Judgment (a:\PCGable121500.JUD)*   Page -84-

Yraguen's Decision- pages 85 thru 104

FINDINGS and CONCLUSIONS:

a. Mr. Abel testified in detail as to how he handled the 30,000 pages of discovery and the steps he took when he initially went through the materials three (3) times. The examination and cross-examination of both the State's and Defense's witnesses revealed that Defense Counsel had in fact read the reports and discovery furnished by the State as well as by their own Defense Investigators. The massive amount of material generated in the form of Defense Notebooks, Defense Summaries, Defense Time Lines, legal research, Defense Investigators' Reports, etc., noted above in the Summaries shows the attention that the Defense Team, including the Defense Attorneys, paid to detailed preparation and planning for the Defense of Mr. Gable. Exhibit #235 is just one example of the sheer mass of materials with which the Defense had to contend and put into some sort of understandable order.

(1) Small examples of attention to detail can at times be indicative and revealing of how the Defense Attorneys were handling reports and discovery furnished by the State. For example, on Defense Motions for Suppression made during the State's Case-in-Chief, the Defense spent considerable time contending that a discovery violation had occurred because the number of pages in the transcription of Frank Gable's statement to Officer Bain was 100 pages in the Defense copy and what appeared to be 102 pages in the one being reviewed by Officer Bain. That statement was a part of the discovery provided to the Defense of document #s 28,281 through 28, 384. See Exhibit #232, Pgs. 7392 through 7400. It eventually turned out that the header had been left off one of copies of the transcript. See Exhibit #232, Pgs. 7440 - 7443. The Defense's motions to compel discovery of the grand jury notes and police officer notes because several police officer reports provided during discovery discussed interviews by persons other than the writer of the reports (See Exhibit #232, Pgs. 43, 50) as well as the Defense's request for the raw data associated with the polygraph tests of witnesses (See Exhibit #232, Pg 95) are other small examples of matters which could only be raised when discovery was being reviewed in some detail.

b. Mr. Gable in his Deposition for these post-conviction relief proceedings stated that his basis for this Claim was because of the letter written by the Defense Investigators after the Depoe Bay meeting discussed in detail in the Summaries above (rather than being able to point to any specific failures). See Exhibit #156, Pg. 35. This record is devoid of any basis for this Claim or of any basis from which even an inference of inattention to the details of the reports and discovery can be discerned. In fact, it appears that serious attention was given to the details of discovery.

c. The transcript of proceedings (Exhibit #232), as per the above Summaries, shows during examination that Defense Counsel had a good command of the information contained in the State's discovery.

CLAIM #5: IT IS ALLEGED THAT DEFENSE COUNSEL ROBERT L. ABEL AND JOHN E. STORKEL FAILED TO READ THE DEFENSE INVESTIGATOR REPORTS, CONSULT WITH THEM ON AN ONGOING BASIS, AND EFFECTIVELY USE THE INFORMATION THE INVESTIGATORS PROVIDED THEM BEFORE AND DURING TRIAL.

FINDINGS and CONCLUSIONS:

a. Although this Claim appears to have been the position of most of the Defense Investigators, as in the summary of their testimony noted above in Mr. Gable's post-conviction case-in-chief, there is little in the way of any specifics as to what was overlooked for purposes of the trial presentation. As previously noted, Mr. Abel stated that he read all of the State's discovery once; then reread the discovery and underlined; and then went back and entered significant portions on his computer program. See Exhibit #3, Pgs. 29. It was from these materials that assignments were given each week to the individual investigators. The most crucial point of alleged failure by the Defense Investigators during the post-

575

1 conviction trial was their general conclusion that Defense Counsel failed to consult with them on an
   ongoing basis. They allege a failure to effectively use the information provided to Defense Counsel
2 without supplying specifics in support of the general claim or showing detriment or prejudice to Mr.
   Gable. It is obvious that there were many many meetings between the Defense Counsel and Defense
3 Investigators as well as the retreat at Depoe Bay. What seems just as obvious is not that there was a
   failure to consult with them, but that the Defense Investigators would have preferred a different and
4 more frequent style of consultation. There has been a complete lack of showing what information that
   was provided by the Defense Investigators was ineffectively used has not been shown.

5
        b. Communication certainly could have been better with Defense Counsel from the Defense
6 Investigators' view, but some of the Defense Investigators seemed to lose sight of the fact that they were
   hired and directed by the Defense Attorneys, and primarily by lead Defense Attorney Robert Abel, and
7 were responsible in turn to the lead Defense Investigator Thomas McCallum. Being part of a Defense
   Team did not imply that pure democratic principles would then operate for all individuals making up
8 the Team. There seemed to be a perception amongst most of the Defense Investigators that given
   enough time there was surely something out there somewhere which would demolish the State's case.
9 The simple fact of the matter was and is, even though they continue to feel as they felt when they wrote
   their letter to Judge West and from the evidence given in this post-conviction trial will probably
10 continue to feel the same way in the future, nothing has surfaced which is any different than
   circumstances that existed and were known at the time of trial. This post-conviction case, after a rather
11 exhaustive review by this Court of the trial transcript of both the guilt and penalty phases, the post-
   conviction trial transcript, reviewing over 250 Exhibits including all sorts of different matters, rather
12 than consisting of anything new, simply appears to have been little more than a reexamination of
   information known at the time of trial which was utilized effectively by both the Defense and the State.
13 This particular Claim is based upon Mr. Gable's assumption that because some of the Defense
   Investigators were dissatisfied with the Defense Team efforts, there was therefore a failure to use
14 information supplied by the State through
   discovery as well as Defense generated materials. Mr. Gable's general assumption is not supported by
15 any specific facts.

16   c. One simply has to review the Summary of the Defendant's case-in-chief set forth above to
   recognize that much of it was generated by the Defense investigation. Defense Counsel would obviously
17 have had to have read the materials generated by the Defense Investigators to have fashioned the defense
   of Mr. Gable which is included in the record of proceedings.

18

19 *CLAIM #6:* IT IS ALLEGED THAT DEFENSE COUNSEL ROBERT L. ABEL AND JOHN E.
   STORKEL, AGAINST ADVICE OF DEFENSE INVESTIGATORS, TURNED OVER ALL
20 INFORMATION OBTAINED BY DEFENSE INVESTIGATORS TO THE STATE AS DISCOVERY
   WITHOUT REVIEWING TO DETERMINE WHICH PORTIONS WERE DISCOVERABLE AND
21 WHICH SHOULD HAVE BEEN KEPT AS NON-DISCOVERABLE WORK PRODUCT OR
   OTHERWISE IRRELEVANT, IMMATERIAL OR NOT DISCOVERABLE.

22
   FINDINGS and CONCLUSIONS:
23
        a. As late as January 11th, 1991, the State's Deputy District Attorneys Thomas Bostwick and
24 Sarah Moore were complaining to the Trial Judge about not receiving reciprocal discovery. The
   complaint was that the State had turned over 27,000 pages of discovery and 1,900 photographs but had
25 received only 60+ pages of discovery from the Defense. See Exhibit #232, Pgs. 177 - 178. Even though
   the Trial Judge emphasized that he expected full reciprocal discovery to be made by the Parties,
26 complaints continued by the Defense and by the State on an ongoing basis regarding the lack of
   discovery. The State continued to complain about the lack of discovery and were continuing to do so
27 even on the eve of the Defendant's case in chief. See Exhibit #232, Pg. 8465.

28 *Frank Edward Gable vs. State of Oregon Post-Conviction Judgment (a:\PCGable121500.JUD)*   Page -86-

1   (1) As an example, on one occasion the State complained about an acknowledged failure to make timely discovery of the defense investigator's notes of a conversation with Greg Allen Johnson
2 until the same afternoon he was called by the Defense, which complaint was acknowledged by the Defense. See Exhibit #232, Pgs. 8629-8630.

3

4   b. Only two specific (2) matters were raised (other than perhaps what Ms. Castle characterizes as one of her confidential memorandums, which appear to be related to this contention): 1) turning over
5 reports dealing with the surveillance that the Defense put on Kevin Francke, the victim's Brother, for a short period of time; and 2) something referred to as possible work-product associated with Jodie
6 Swearingen. As previously noted, Mr. Abel denies that either the surveillance reports on Kevin Francke or the memorandums were turned over to the State. See Exhibit #3, Pgs. 33 and 34. On the second
7 point regarding Jodie Swearingen, it is recognized that the material would have to have been turned over to the State, but the contention seems to be that it didn't need to be turned over to the State as early as
8 it was provided to the State.

9   (1) Jodie Swearingen was one of the State's material witness, who was even held in custody for a considerable period of time. The Defense material, such as that involving Jodie
10 Swearingen, was required to be turned over forthwith, and the Defense agreed to do so very early in the trial proceedings. See Exhibit #232, Pgs. 209 through 221. The contention that the Swearingen material
11 didn't have to be turned over to the State as early as it was provided is nonsense.

12   c. It appears that some of the major problems between Defense Counsel and the Defense Investigators were who was going to control the determination of what was going to be investigated as
13 well as what was going to be turned over to the State in the form of discovery and when such discovery was going to be made. This Court would conclude, from the testimony of Robert Abel and the Affidavit
14 of Grace Castle, of which this Court is somewhat dubious considering the circumstances which led to its preparation, that it was the State which was coming out on the short end of the bilateral discovery
15 process, rather than the Defense giving up unnecessary information. It was the Defense Counsel who forced a good faith compliance with discovery apparently to the chagrin of some of the Defense
16 Investigators. The Defense Investigators had no business even making this a point of contention. Such a position by the Defense Investigators is hardly the basis for a meaningful post-conviction relief Claim.

17   d. As a side note, it was also of interest that the Defense apparently did not divulge information on at least one individual subpoenaed by Defense Investigator Nesbitt-Lance, who was told during the
18 penalty phase to get the potential witness away from the Courthouse, because he was a witness who Mr. Abel in his Deposition says "was walking across the grounds in a big, open field, I think there's a
19 Hawthorne Street or something by the dome, and saw this guy running away from the Dome Building about the time of the murder, and he describes a guy that fits Frank Gable to a T." See Exhibit #3, Pgs.
20 17 and 18. This may be the one exception to this Court's previous statement that no new information was uncovered during the review conducted in these post-conviction relief proceedings.

21

22 CLAIM #7: IT IS ALLEGED THAT DEFENSE COUNSEL ROBERT L. ABLE AND JOHN E. STORKEL FAILED TO EFFECTIVELY CROSS-EXAMINE THE STATE'S WITNESSES IN
23 GENERAL, AND THE FOLLOWING IN PARTICULAR:

24

      a. JODIE SWEARINGEN:
25

  FINDINGS and CONCLUSIONS:
26

      a. Jodie Swearingen was not called by the State, but rather by the Defense in its case in chief.
27 See Exhibit #232, Pgs. 9324 through 9371. As noted in the summary of the Defendant's case, Ms.

28

1 Swearingen disavowed all of the other statements she had given to law enforcement officers as well as at least five (5) other individuals and, in essence, testified that on the evening of January 17th, 1989, she
2 was not the grounds of the Dome Building, that she did not see Mr. Gable burglarizing a car, and that she did not see Mr. Gable stab Mr. Francke. See Exhibit #232, See Pg. 9329.

3

    b. The Defense was obviously prepared to impeach Ms. Swearingen [See PCT, Vol. I, Pgs. 143,
4 144], but such was unnecessary in view of the nature of Ms. Swearingen's testimony.

5   c. The main thrust of Mr. Gable's Claim appears to be a contention that the Defense should have impeached their own witness through other statements that she made followed by other testimony, as
6 for example through her Father, that she was in Dundee at about 6:30pm on January 17th, 1989. In Mr. Gable's view this would impeach her but then indirectly impeach the testimony of Cappie Harden, who
7 the State did call and who did testify that he, accompanied by Jodie Swearingen, observed Frank Gable stab Michael Francke.

8

9   (I) As noted in the evidence Summary, any attempted impeachment would not have been accomplished as contemplated by Mr. Gable, even if attempted. Joe Swearingen, for example, would
10 not have added much to the proceedings. Pursuing such impeachment would have been counterproductive to what the Defense had obtained on Jodie Swearingen's direct examination. Mr.
11 Gable's approach to Ms. Swearingen's evidence is as this Court has previously noted – rather convoluted.

12

    b. CAPPIE HARDEN:
13

FINDINGS and CONCLUSIONS:
14

15   a. As noted in the Findings and Conclusions above, Defense Counsel John E. Storkel cross-examined Cappie Clifford Harden, who was presented during the State's case in chief, for one-hundred
16 sixty-seven (167) pages of transcript [See Exhibit #232, Pgs. 8078 through 8145]. The transcript reveals a series of inconsistencies in the testimony of Cappie Harden covering his interviews with law
17 enforcement authorities on November 20th, 1989, January 18th, 1990, January 20th, 1990, January 21st, 1990, and January 24th, 1990. The inconsistences included, but were not limited to: the color of the
18 Gable Toyota; where Gable left him off during the alleged ride in the vehicle and whether he gave Frank Gable money for the ride; when he met Frank Gable and, in particular, whether it was after the Francke
19 murder; the circumstances of getting Jodie Swearingen back from Dundee; what he and Jodie Swearingen did after she returned to Salem; the nature of his relationship with Jodie Swearingen and
20 whether he did or did not know her age; his knife collection and whether he had sold or given away knives; what he had heard on the streets from others about the Michael Francke killing and whether he
21 was just giving their accounts while inserting himself; about his personal use of drugs and the effect upon him; how familiar he was with the State Hospital grounds including how many times he did or did
22 not drive past the State Hospital grounds; not being able to precisely locate where he was parked on the parking lot of the Dome Building; not being able to judge the distance he was from the Francke vehicle
23 when he was parked on the parking lot of the Dome Building; the fact that he thought Frank Gable was a rat and had a low opinion of him; whether he had or had not seen Frank Gable earlier in the day; the
24 nature of the deals he had made with the District Attorney's Office involving crimes charged against him; getting Cappie Harden to acknowledge that he had lied about everything associated with his
25 statements to the police at one time or another; and the details of the letter he wrote to his Nephew Richard Swain about whether he should fry the rat Frank Gable and take the freedom and cash his
26 testimony would bring. It is obvious that John Storkel had done a meticulous review of the statements and reports concerning Cappie Harden and did pursue detailed matters for page after page of transcript.

27   b. Mr. Gable has prepared a Notebook entitled "Impeachment of Cappie Hardin" (See Exhibit

28 *Frank Edward Gable vs. State of Oregon Post-Conviction Judgment (a:\PCGable121500.JUD)*   Page -88-

1 #12). However, as is noted above, the Defense did an excellent job of cross-examining Cappie Harden, and it does not appear that any new information for impeachment purposes than was covered by the
2 Defense *has* been included in the Notebook.

3

c. JANYNE GABLE:

4

FINDINGS and CONCLUSIONS:

5

a. Janyne Gable was presented as a witness in the State's case in chief. The Defense, at the
6 conclusion of her direct examination, simply indicated that they did not have any questions for Janyne
Gable at that time. See Exhibit #232, Pg. 7733.

7

b. Mr. Abel testified that he reinterviewed Janyne Gable after she had testified, and he decided
8 against recalling her because it was apparent that she was going to remain firm in her testimony, and he
did not want to simply reemphasize her testimony. See PCT, Vol. III, Pgs. 522, 539. The explanation
9 appears to be plausible.

10 c. Mr. Gable's position regarding matters which he contends should have been used to impeach
Janyne Gable involve matters associated with the "Gable Time Line" (See Exhibit #9), that is, the
11 proposed alibi defense, and a report (See Exhibit #192) from Defense Investigator Hamilton involving
an interview with Michael F. Goss, who had an ongoing relationship with Mr. Gable since prison, and
12 who stated it was possible he might have spent the night at Mr. Gable's residence. The report says that
Mr. Goss says he recalled Janyne Gable getting a telephone call from a hospital employee that Michael
13 Francke had been murdered. The inclusion of the material in this notebook would imply that Janyne
Gable should have been cross-examined for potential impeachment testimony from Mr. Goss. However,
14 Mr. Goss did not state that he was definitely at the Gable house the night of January 17th, 1989, and
pursuing such a matter with Janyne Gable could have in fact been detrimental to the Defense. It appears
15 to have been a tactical decision, which certainly as in the case of most tactical decisions can be
subsequently questioned, but the materials presented by Mr. Gable are not such that a conclusion can
16 be drawn that if Janyne Gable had been cross-examined that more benefit could have been gained then
detriment or that, in any event, it would have had some sort of bearing on the trial outcome. From the
17 materials reviewed, it would be this Court's conclusion that Mr. Abel probably made the right tactical
decision in not putting Mr. Goss on the stand.

18

19 d. MIKE KEERINS:

20 *FINDINGS and CONCLUSIONS:*

21 a. Mike Keerins was not called by the State or the Defense and thus was not subject to cross-
examination. However, on this point, it should be noted that Kenneth Hadley, in his opening statement
22 for the post-conviction proceedings, said that there were some mistakes made in the names listed under
this category. See PCT, Vol. I, Pg. 21. The fact that this was a nonissue was further reiterated during
23 trial. See PCT, Vol. II, Pgs. 328, 329.

24 b. In materials submitted by the State, it is of interest that in investigative reports prepared by
defense investigators, Mike Keerins is saying that Frank Gable told him that he was at the car when
25 Michael Francke came up and tried to hold him for the cops and that he stabbed the guy. See Exhibit
#s 172 and 173.

26

27 e. JOHN CROUSE:

28 *Frank Edward Gable vs. State of Oregon Post-Conviction Judgment (a:\PCGable121500.JUD)*   Page -89-

579

1 *FINDINGS and CONCLUSIONS:*

2   a. Johnny Lee Crouse was not called by the State and thus not subject to cross-examination by
   the Defense.

3

4   b. Johnny Crouse was called by the Defense on June 17th, 1991, and subjected to direct
   examination by Mr. Abel. See Exhibit #232, Pgs. 9486 through 9489. Johnny Lee Crouse exercised his
   Fifth Amendment rights and refused to answer all questions except in response to the initial question

5 "In January 17, 1989, did you kill Michael Francke?, he answered "No". See Exhibit #232, Pg. 9487.
   After extended colloquy, a recalling of Mr. Crouse by the Defense [See Exhibit #232, Pgs. 9515, 9516],

6 taking testimony from Oregon State Police Officer Randy C. Martinek, receiving tape recordings and
   transcripts of law enforcement contacts with Johnny Crouse, and stipulations involving the State's

7 efforts to investigate Johnny Crouse's possible involvement in the murder of Michael Francke through
   an extended offer of proof which lasted through noon on June 18th, 1991, the Court excluded the

8 testimony of Mr. Crouse on the basis of "irrelevancy". See Exhibit #232, Pg. 9511.

9   c. The Defense went to great lengths to procure the presence of Mr. Crouse, subject him to
   examination and present his testimony before the Jury. The Defense preserved for the record its

10 exception to the exclusion of the evidence involving Mr. Crouse.

11   d. This subsection was also noted by Mr. Hadley as being a nonissue for purposes of the post-
   conviction relief proceeding. See PCT, Vol. II, Pgs. 328, 329.

12

13   f. JOHN KEVIN WALKER:

14 *FINDINGS and CONCLUSIONS:*

15   a. Defense Counsel Robert Abel cross-examined John Walker, who was presented as a witness
   in the State's case in chief, for twelve (12) [See Exhibit #232, Pgs. 8187 through 8199] pages of

16 transcript. The cross-examination centered on circumstances about Mr. Walker being a snitch which
   brought about the attack upon him in prison; the fact that he was a major drug seller making between

17 $300 to $400 per day; that Mr. Walker was a heavy user of methamphetamine during this time period
   and had used crank on both January 17th and January 18th, 1989; and that fact that he was a snitch, even

18 though he claimed not to be, and why should anyone trust the word of a snitch. Mr. Hadley also noted
   this subsection as being a nonissue for purposes of the post-conviction relief proceeding. See PCT, Vol.

19 II, Pg. 329.

20   b. The cross-examination by Mr. Abel of John Walker was effective and competently done.

21

22   g. KRIS KEERINS:

23 *FINDINGS and CONCLUSIONS:*

24   a. Kris Keerins was not called by the State or the Defense and thus not subject to cross-
   examination.

25   b. However, on this point, it should be noted that Kenneth Hadley, in his opening statement for
   the post-conviction proceedings, said that there were some mistakes made in the names listed under this

26 category. See PCT, Vol. I, Pg. 21. The fact that this was a nonissue was further reiterated during trial.
   See PCT, Vol. II, Pgs. 328, 329.

27

c. Of unrelated interest was that Mr. Gable in his Deposition stated that the Keerins Brothers, Mike and Kris, knew information about Tim Natividad and Johnny Crouse being involved in the murder of Michael Francke, which is discussed under the next Claim, which information could be confirmed by Jodie Swearingen. See Exhibit #156, Pgs. 33, 38 through 40. It appears that Mr. Gable was suggesting that these witnesses should be called for such purpose. This is the same Mike Keerins who allegedly was told by Mr. Gable that he had stabbed Michael Francke.

Finally, with respect to the witnesses named in this particular Claim, it was surprising to this Court in reviewing the trial transcript that Mr. Gable wasn't more concerned about the cross-examination of witnesses such as Linda M. Perkins, Mark Gesner, Earle Childers and David Walsh, all of whom testified that Mr. Gable admitted the murder of Michael Francke, who in this Court's view, gave far more damaging testimony against Mr. Gable than some of the witnesses named above, rather than some of the above-named witnesses, a number of which became nonissues.

CLAIM #8: IT IS ALLEGED THAT DEFENSE COUNSEL ROBERT L. ABEL AND JOHN E. STORKEL FAILED TO ADEQUATELY INVESTIGATE TIMOTHY NATIVIDAD'S AND/OR JOHN CROUSE'S INVOLVEMENT IN THE MURDER OF MICHAEL FRANCKE.

FINDINGS and CONCLUSIONS:

a. With respect to the Timothy Natividad matter, Petitioner Frank Gable's first appearance with Counsel Abel and Storkel was on April 12th, 1990. See Exhibit #235, Vol. 1, Pg. 18.

(1) By August 28th, 1990, Defense Counsel, Mr. Able and Mr. Storkel, were actively pursuing matters involving Timothy Natividad as a potential suspect in the murder of Michael Francke in Court proceedings. See Exhibit #235, Vol. 4, Pgs. 72 through 81.

(2) On September 12th, 1990, Mr. Able and Mr. Storkel were again pursuing the Natividad matter with the Court [See Exhibit #235, Vol. 5, Pgs. 98 and 99]; and again on October 3rd, 1990 [See Exhibit #235, Vol.6, Pgs. 108 through 113]; and again on November 5th, 1990 [See Exhibit #235, Vol. 7, Pgs. 141 and 142]; and again on January 11th, 1991 [See Exhibit #235, Vol. 8, Pg.181 and 190; and again on January 23rd, 1991 [See Exhibit #235, Vol. 9, Pgs. 199 and 200.

(3) The matter was raised by the State in pretrial matters heard on April 30th, 1991, seeking reciprocal discovery regarding the Natividad matter, and it is apparent from the colloquy that the State was aware that the Defense had their pathologist, Dr. Spitz, and their criminalist, Mr. Fox, examine knives which had belonged to Timothy Natividad and in a motion in limine was seeking to limit that Defense from pursuing the subject in front of the jury. See Exhibit #235, Vol. 44, Pgs. 5836, 5837 and thereafter. It further appears from the colloquy that the State was concerned about the jury being contaminated through improper hearsay involving another alleged perpetrator and wanted a preliminary ruling requiring any matter involving Natividad out the presence of the jury [See Pgs. 5852 and 5853]. Mr. Abel advised the Court that the Defense was still investigating the possible tie involving Timothy Natividad, and the Court simply emphasized that if and when there was "admissible" evidence that someone else committed the murder, he would again take up the issue [See Pgs. 5857 and 5858].

(4) The State was anticipating that the Defense would continue to attempt to pursue the Timothy David Natividad matter as it concluded its case in chief, and again raised the matter in a motion in limine. See Exhibit #232, Pg. 8466.

(5) When the matter subsequently came up later, the Court again reiterated that before the Court would allow the matter to be raised before the jury, the Defense would either have to lay a

1 proper foundation or put on evidence that would make the contention regarding Timothy Natividad
   relevant. See Exhibit #232, Pg. 9315.

2

3   (6) It is apparent from the transcript of the trial proceedings that considerable time and
   effort was spent by the Defense investigation team as well as by Defense Counsel in pursuing any lead
   which might establish Timothy Natividad as a potential suspect.

4

5   (7) It is further apparent from the transcript in these post-conviction proceedings, as
   summarized above and in particular with respect to their contentions involving Timothy Natividad, and
   a close analysis of the exact basis for the suspicions of Kevin B. Francke and Elizabeth Francke fkn

6 Godlove regarding their belief that Timothy Natividad's was involved in the death of Michael Francke,
   the specifics do not even begin to rise to the level of any sort of legally recognized reasonable suspicion

7 or probable cause.

8   (i) It is recognized that some of the State's Exhibits do contain reports which
   includes hearsay information that Timothy Natividad was responsible for the death of Michael Francke.

9 However, these proceedings provided the opportunity to come forth with whatever evidence there might
   have been or might be regarding the accusation against Timothy Natividad, and such evidence is simply

10 lacking.

11   (8) It is of interest that Mr. Gable on the one hand complains about Mr. Abel's pursuit
   of his conspiracy theory as noted in the Summaries and yet on the other hand contends under this Claim

12 that the Timothy Natividad matter should have been more adequately investigated. At times it appears
   that Mr. Gable is simply pursuing matters at the same time in different directions.

13

14   b. With respect to John Lee Crouse, Defense Counsel Abel was pursuing the matter with respect
   to the State's motion in limine at the hearing of April 30th, 1991 [See Exhibit #235, Vol. 44, Pgs. 5847

15 and thereafter]. Obviously the Defense was actively investigating the matter which caused the State to
   anticipate that the Defense would attempt to continue to pursue matters involving the confessions of
   John Lee Crouse and thus addressed the matter in a motion in limine before the Defendant started to

16 present evidence. See Exhibit #232, Pg. 8466. The Defense advised the Court that it did intend to
   present the testimony of Johnny Lee Crouse, had Mr. Crouse under subpoena, and that on June 11th,

17 1991, Mr. Crouse had absented himself. See Exhibit #232, Pgs. 8643, 8644, 8715 and 8716, 8986. That
   in turn led to a lengthy hearing on June 12th, 1991, during which the Defense requested a warrant for the

18 apprehension of Mr. Crouse. See Exhibit #232, Pgs. 9106 through 9160.

19   (1) At the June 12th, 1991, hearing the Defense called Sgt. Julian Rodriquez, an Oregon
   State Penitentiary Corrections Officer, who gave the subpoena and instruction letter to Mr. Crouse; Mark

20 Hamilton Jensky, a process server, who delivered the paperwork to Sgt. Rodriquez; Thomas McCallum,
   Tom Nisbet-Lance, and Roger Keith Harris, Defense Investigators, who gave testimony regarding their

21 efforts during the two weeks preceding the June 12th hearing to find Mr. Crouse. The testimony revealed
   that Mr. Crouse was treated somewhat differently than many of the other witnesses that the Defense had

22 under subpoena, and although he was to immediately report after his release on March 25th, 1991, he did
   not do so and the Defense made no effort to contact or find him until investigators were sent to Bend,

23 Oregon, to attempt to find Mr. Crouse.

24   (2) The Court issued a warrant for the apprehension of Mr. Crouse on June 13th, 1991.
   See Exhibit #232, Pg. 9389.

25

26   (3) Johnny Lee Crouse was apprehended and was produced in Court on June 17th and
   18th, 1991. In an extensive offer of proof [See Exhibit #232, Pgs. 9477 through 9567, the Defense
   expended substantial time and effort in an attempt to get the testimony of Mr. Crouse before the jury.

27 See also Findings and Conclusions under 7e under this First Claim. As noted above, other than stating

28 *Frank Edward Gable vs. State of Oregon Post-Conviction Judgment (a:\PCGable121500.JUD)*   Page -92-

1 that he did not murder Michael Francke, Mr. Crouse exercised his Fifth Amendment rights against self-
   incrimination. In February of 1989, Mr. Crouse first stated that he saw five (5) individuals beating on
2 Michael Francke, one of whom was a Hispanic who Mr. Crouse chased to Market Street. See Exhibit
   #232, Pg. 9535. On April 4th, 1989, Mr. Crouse told of being approached by "Juan", who was wearing
3 a three-piece suit, who offered him $300,000 to kill Michael Francke; that he did stab Michael Francke
   and ran East towards Park Avenue, then back pushing Michael Francke against his car before running
4 to the West. See Exhibit #232, Pgs. 9537, 9538, 9540, 9541. On April 5th, 1989, Johnny Crouse had
   changed his story relating that he had used a piece of wire to gain entry to the Francke vehicle,
5 confronted Michael Francke because he wanted his records due to being tired of not feeling out of jail,
   and that he stabbed Michael Francke several times, including his stomach. See Exhibit #232, Pgs. 9543
6 through 9545. By April 11th, 1989 he was recanting his confessions; then on 4/13/89 saying he did it;
   and then on 11/30/89 again denying the killing. See Exhibit #232, Pgs. 9547 through 9549. Although
7 there were some consistencies with what had happened [See Exhibit #232, Pgs. 9550, 9551], there were
   many many inconsistencies [See Exhibit #232, Pgs. 9552 through 9558] and State Police follow-up
8 investigations which produced no results [See Exhibit #232, Pgs. 9564 through 9566].
           (4) Defense Counsel not only fully investigated the potential involvement of Johnny
9 Crouse, they made a herculean effort to get the testimony of Johnny Crouse before the Jury during the
   guilt phase of the trial.
10
           (a) The Defense, to wit, Mr. Abel, did manage to raise the matter of John
11 Crouse's confession in cross-examination of Oregon State Police Paul Bain. See Exhibit #232, Pg.
   7478.
12
           (b) See also Exhibit #209, Motion to Allow Johnny Lee Crouse Evidence, and
13 #210, a Defense Summary of materials reviewed regarding Johnny Crouse.
14
           (5) Even during the penalty phase of the trial, Mr. Storkel was still attempting to get
   matters into evidence regarding Johnny Crouse, and as related below, matters involving Jose Navarro.
15 See Exhibit #232, Pg. 10470
16   c. Defense Counsel Able and Storkel also pursued other possible suspects such as:
17   (1) Samuel Rosillo Cornejo aka Jose Chicua who from information passed by a
   confidential police informant, had allegedly told Angela Myer and Dorothy McDonald that he had killed
18 Michael Francke [See Exhibit #232, Pgs. 5829 through 5836];
19   (2) Jose Navarro by serving a subpoena duces tecum on Dr. Peter Batten in an attempt
   to obtain the Navarro autopsy report [See Exhibit #232, Pgs. 8515-8516].
20
   d. The Timothy Natividad and Johnny Crouse aspects were pursued and beaten to death at trial,
21 and although truly having no life beyond speculation and conjecture, even at this point they are
   continuing to be beaten on in these post-conviction proceedings to little avail.
22
23 CLAIM #9: IT IS ALLEGED THAT DEFENSE COUNSEL ROBERT L. ABLE AND JOHN E.
   STORKEL FAILED TO ADEQUATELY DEVELOP, INVESTIGATE AND PRODUCE AT TRIAL
24 EVIDENCE THAT TIMOTHY NATIVIDAD AND/OR JOHN CROUSE WAS THE KILLER OF
   MICHAEL FRANCKE.
25
26 FINDINGS and CONCLUSIONS:
27   a. The matters covered in the detailed discussion of Issue #8 above fully negate this contention.
28 *Frank Edward Gable vs. State of Oregon Post-Conviction Judgment (a:\PCGable121500.JUD)*   Page -93-

CLAIM #10: IT IS ALLEGED THAT DEFENSE COUNSEL ROBERT L. ABLE AND JOHN E. STORKEL FAILED TO ADEQUATELY DEVELOP, INVESTIGATE AND PRODUCE A QUALIFIED EXPERT TO CONDUCT AN EXAMINATION OF THE AUTOMOBILE DRIVEN BY TIMOTHY NATIVIDAD ON THE NIGHT OF THE DEATH OF MICHAEL FRANCKE, FOR BLOOD OR OTHER TRACE EVIDENCE THAT WOULD HAVE CONNECTED HIM TO THE KILLING.

FINDINGS and CONCLUSIONS:

a. Nothing has been established in a review of the trial transcript or post-conviction transcript or Exhibits received in this case to establish that the Timothy Natividad vehicle was related to the murder of Michael Francke or that it should have been checked for blood. The totality of this Claim appears to be based upon the reasoning of Kevin Francke, the victim's brother, as set forth in the summary of his testimony in the post-conviction trial. If adequate evidence could not be established to connect Timothy Natividad with the murder of Michael Francke, then how can it be error for the Defense, having exhausted all leads related to Timothy Natividad, to have failed to somehow seize, presumably by Court order, or obtain the Natividad vehicle (assuming there was only one possible vehicle which the evidence does not seem to necessarily support) and then to obtain Court authorization for funds to pay a qualified expert to examine that vehicle?

b. The Natividad vehicle matter is simply something that Mr. Gable says should have been done without any specific basis other than the general conclusion that the real murderer of Michael Francke was Timothy Natividad. No basis has been established for this Claim.

CLAIM #11: IT IS ALLEGED THAT DEFENSE COUNSEL ROBERT L. ABLE AND JOHN E. STORKEL FAILED TO SUBPOENA AND PRESENT TESTIMONY AT THE TRIAL OF THE WIFE, FAMILY AND OTHERS ASSOCIATED WITH TIMOTHY NATIVIDAD THAT WOULD HAVE SHOWN MR. NATIVIDAD WAS THE KILLER OF MICHAEL FRANCKE AND NOT FRANK GABLE.

FINDINGS and CONCLUSIONS:

a. See notes under Claims #8, #9 and #10 above regarding how the Defense handled matters involving Timothy Natividad. Lisa Godlove Francke, the "wife" of Timothy Natividad, was called in these post-conviction relief proceedings. It is doubtful that her testimony or the testimony of any of the Natividad Family members would have been admissible at trial. A review of Lisa Franke's testimony is set forth in the post-conviction summary of Mr. Gable's evidence. She, and her Husband Kevin Francke, and even Charlie Burt, Lisa's attorney in her murder trial if Lisa's representations of Mr. Burt's position are accurate, are certainly entitled to believe whatever they want to believe – the only difficulty in this case is that there wasn't sufficient evidence for even a probable cause finding against Timothy Natividad, let alone an indictment and prosecution for the murder of Michael Francke.

CLAIM #12: IT IS ALLEGED THAT DEFENSE COUNSEL ROBERT L. ABEL AND JOHN E. STORKEL FAILED TO ADEQUATELY REPRESENT THE DEFENDANT BY ROBERT L. ABLE ENGAGING IN A PATTERN OF EXCESSIVE CONSUMPTION OF ALCOHOL DURING PREPARATION FOR THE TRIAL AND DURING THE TRIAL.

FINDINGS and CONCLUSIONS:

a. Robert Abel did consume alcoholic beverages at social occasions during the period of time

1   he represented Frank Gable.

2   b. It is also apparent that Mr. Abel's consumption of alcoholic beverages became excessive during a time subsequent to Mr. Gable's trial which caused Mr. Able to enter an alcoholic abuse
3   treatment program for twenty days in 1994.

4   c. As per the summaries, it is apparent that there is a wide divergence about whether Mr. Abel did or did not ever have the odor of alcohol about his person during the Frank Gable trial. Mr. Abel
5   testified that he did not drink before trial proceedings would begin or during trial proceedings. If there was any odor, the odor was not as bad or as apparent as has characterized by Mr. Gable and two of his
6   witnesses.

7   d. Most importantly, the consumption of alcoholic beverages during that period of time and/or the odor associated with such beverages has not been shown to have had any detrimental or prejudicial
8   effect upon the trial in this case.

9   e. This Court views this particular Claim as simply being an attack on Mr. Abel's character, first raised in Mr. Gable's letter between the guilt and penalty phases, which is not supported by the Trial
10   Court or other officers of the court, whose business it would have been to be attentive to such matters. More particularly, is not supported by the record when the performance of Mr. Abel during trial as
11   reflected in the transcript of proceedings is reviewed in detail. Mr. Abel demonstrated on the record that his abilities as a criminal defense attorney are varied and considerable.

12

13   *CLAIM #13:* IT IS ALLEGED THAT DEFENSE COUNSEL ROBERT L. ABEL AND JOHN E. STORKEL FAILED TO PROPERLY OBJECT AND ARGUE TO THE TRIAL COURT THAT THE
14   INDICTMENT REFERRED TO ABOVE WAS INVALID BECAUSE THE GRAND JURY FOREMAN, THOMAS H. DENNEY, OSB NO. 66034, WAS A CAREER PROSECUTOR
15   EMPLOYED BY THE DEPARTMENT OF JUSTICE OF THE STATE OF OREGON AND UNAUTHORIZED PERSONS WERE PRESENT AT THE GRAND JURY.

16

FINDINGS and CONCLUSIONS:

17   a. This matter was actually raised formally by Defense Counsel in Motion to Quash and/or
18   Resubmit the Indictment to the Grand Jury on the basis of a professional and collegial relationship between the Foreman, Thomas H. Denney, and Scott McAlister, an individual investigated as a part of
19   the case, which was filed on July 20th, 1990, to which the State responded with a Memorandum in Opposition filed August 28th, 1990. See Exhibit #s 103 and 104. The basis of the Motion was bias since
20   both Thomas Denney and Scott McAlister were Assistant Attorney Generals for the State of Oregon. Although this basis is not included within statutory grounds for setting aside an indictment set forth in
21   ORS 135.510, an argument could have been advanced, if in fact bias could be shown which was not done in this case, on Constitutional grounds. On October 11th, 1990, the Defense in writing elected not
22   to pursue the said Motion to Quash and/or Resubmit. See Exhibit #105.

23   (1) The reasoning is explained in Mr. Storkel's Deposition in which he states that no chicanery, conspiracy or corruption could be established (See Exhibit #2, Pgs. 40 and 57) with the only
24   remaining aspect being simply "it didn't look good" (See Exhibit #2, Pg. 40).

25   (2) There was simply no statutory, Constitutional, or factual basis on which to proceed with the Motion based on a professional and collegial relationship with Scott McAlister..

26

27   b. As evidence in the post-conviction trial showed, there were actually several persons in the Department of Justice who had limited contact or involvement with the investigation and prosecution

28   *Frank Edward Gable vs. State of Oregon Post-Conviction Judgment (a:\PCGable121500.JUD)*   Page -95-

1 of the Gable case including Brenda Peterson, Randy Martinek, Katherine McLaughlin, Chuck Pritchard and Bob Hamilton. None of these individuals have been shown to have had any contact with Thomas
2 Denney during his involvement in the Grand Jury investigation of the Michael Francke murder. In like fashion, the involvement of the Department of Justice in the original investigation and prosecution do
3 not provide a basis for a collateral attack upon Thomas Denney from carrying out his duties as the Foreperson of the Grand Jury.

4

     c. Circuit Judge Duane R. Estagaard signed an Order on September 5th, 1989, which Order was
5 filed on September 6th, 1989, allowing William J. Pierce to be allowed to appear as a special agent for the Grand Jury. See Exhibit #101. Although it does not appear from the record that Defense Counsel
6 objected to this Order or the appearance resulting therefrom, no showing has been made in these proceedings that the Order and/or appearance was improper or illegal.

7

8 *CLAIM #14:* IT IS ALLEGED THAT DEFENSE COUNSEL ROBERT L. ABEL AND JOHN E. STORKEL FAILED TO OBJECT ON THE GROUNDS OF EX POST FACTO THE COURT'S
9 SUBMITTING TO THE JURY IN THE PENALTY PHASE OF THE TRIAL THE POSSIBILITY OF THE DEFENDANT BEING SENTENCED TO LIFE WITHOUT THE POSSIBILITY OF PAROLE.

10

FINDINGS and CONCLUSIONS:

11

     a. This is the one issue in all of the Claims in this post-conviction relief proceeding, which has
12 presented, in this Court's view, the most difficulty in resolution. The issue arises in this case because the true life option did not become effective until after the murder of Michael Francke. At the time of
13 the murder of Michael Francke the options available in the penalty phase after conviction of aggravated murder included the death penalty or imprisonment with the possibility of parole. It seems clear from
14 recent case law that a defendant facing the imposition of the death penalty can waive or relinquish the protection of the State and Federal *ex post facto* Constitutional protections against a retroactive
15 application of the true life option. See *State vs. Langley*, Or (2000 – SC S41885 filed 12/29/00); *State vs. Rogers*, 330 Or 282 (2000); *State vs. McDonnell*, 239 Or 375 (1999). The true life option
16 affords an additional alternative to a jury considering a death penalty case short of imposing the death penalty, particularly when a jury may be of an opinion that the convicted defendant should never again
17 be released into society. This issue in this case narrows to whether there was or was not a waiver or relinquishment of the State and Federal *ex post facto* Constitutional protections. In this case, Mr. Gable
18 did not waive or relinquish the *ex post facto* protections in writing or verbally on the record. Mr. Gable in this Claim contends that there is thus a silent record and a waiver or relinquishment can never be
19 implied from a silent record. However, the record is not silent regarding the *ex post facto* issue.

20   (1) The Defense during the initial stages of potential juror voir dire examination made reference to the true life option. When the legal *ex post facto* issue was placed on the record by Deputy
21 District Attorney Thomas Bostwick on April 1st, 1991, Mr. Storkel told the Court that the issue would be discussed with Mr. Gable. At the Court's conclusion on the record was, however, that the Parties
22 were in agreement that the true life option could be discussed during voir dire proceedings to which both acceded. On April 2nd, 1991, Mr. Able told the Court that the matter had still not been discussed with
23 Mr. Gable, and that the Defense would not be addressing that issue with the prospective jurors. Mr. Gable was present during the proceedings on both April 1st, 1991, and April 2nd, 1991.

24

25   (2) The Trial Judge Greg West testified that the matter was discussed, perhaps in chambers, and his impression was that the choice of the Defense was to go with the life without parole
26 as a defense strategy to give jurors a viable choice other than the death penalty or life with parole. This assumption involving the defense strategy which included the life without parole option would be
27 reinforced by the fact that, as noted in the Summary of proceedings above:

28

1   (i) The Defense specifically requested the life without parole option as a written requested Jury Instruction;

2

3   (ii) The Court prepared proposed Jury Instructions incorporating the life without parole option which he presented to the Parties the afternoon before giving preliminary instructions to the Jury at the beginning of the Penalty Phase of the trial. These same provisions given in the preliminary instructions, including the three options, that is: 1) the death penalty, 2) the life without parole penalty, and 3) the life with parole penalty, were included in the Court's Final Jury Instructions.

4

5   (iii) Mr. Gable upon the return of the jury verdict waived time and requested that he be sentenced immediately. Everyone involved, that is, the Court, the State and the Defense had no question from the Jury's Answers to the Verdict Questions that the result was the life without parole penalty. Mr. Gable, from the beginning when the issue was first raised during jury selection, had the right to assert the *ex post facto* and never asserted such right. In fact, all of the actions which are reflected on the record are to the contrary.

6

7

8

9   b. It is this Court's finding that there was a waiver of the *ex post facto* protections which can be implied from the totality of the circumstances reflected in the record of this case. The Defense strategy was obviously effective in that the Jury selected the option, life without parole. Mr. Storkel's impression that the Jury might have gone with the death penalty option rather than the life with parole option if only those two choices had been presented may be correct which would lead to the conclusion that Mr. Gable was greatly benefitted by the Defense strategy. In any event, there is nothing in the record to show that the Jury seriously considered the life with parole option which Mr. Gable is now seeking to obtain under this Claim.

10

11

12

13

14   **SECOND CLAIM OF RELIEF**
         PETITIONER GABLE REALLEGES SECTIONS 1 THROUGH 8 OF HIS FIRST CLAIM AND ALLEGES THAT HE WAS DENIED THE RIGHT TO TESTIFY ON HIS OWN BEHALF UNDER THE FIFTH AND FOURTEENTH AMENDMENTS TO THE UNITED STATES CONSTITUTION AND UNDER ARTICLE I, SECTION 11, OF THE OREGON STATE CONSTITUTION IN THAT DEFENSE COUNSEL FAILED TO ALLOW HIM TO TESTIFY IN HIS OWN BEHALF IN THE GUILT PHASES PORTION OF HIS TRIAL KNOWING THAT HE WAS UNWILLING TO WAIVE HIS RIGHT TO TESTIFY IN HIS OWN BEHALF.

15

16

17

18

FINDINGS and CONCLUSIONS:

19

20   a. As noted above in a the Summary of post-conviction proceedings, Paul A. Dahalopos testified that he was hired by the Defense to prepare Frank Gable for cross-examination by the State in the event he was called as a witness in his own behalf. However, Mr. McCallum, a defense investigator, advised him that the mock cross-examination session set for May 17th, 1991, was to be postponed, and it was not rescheduled. In putting this matter in perspective, one needs to recognize that the postponement occurred in the middle of the State's case-in-chief which began with the State's opening statement on May 1st, 1991, and ended on May 31st, 1991, with the Defense not even beginning to present evidence until June 5th, 1991.

21

22

23

24   b. Some deductions from the Summary seem apparent: 1) The Defense Investigators, Mr. Gable and Ms. Steele were all aware that Defense Counsel did not believe that Mr. Gable should testify; 2) Mr. Gable had not decided whether he was or was not going to testify throughout trial; i.e., he was equivocating waiting to see what was going to happen on the Defense's case-in-chief; 3) Mr. Abel assumed that Mr. Gable was in agreement with his conclusion that he should not testify, for reasons outlined by Mr. Abel, when he rested the Defense; 4) Mr. Gable never told Mr. Abel or Mr. Storkel that he was going to testify; 5) Even Mr. Gable's letter to the Judge West between the Guilt and Penalty

25

26

27

28   *Frank Edward Gable vs. State of Oregon Post-Conviction Judgment (a:\PCGable121500.JUD)*   Page -97-

1 Phases does not state that he demanded to testify and was denied that right, but rather that he was denied his right that numerous witnesses testify on his behalf which denial "effectively took away my option

2 to testify on my own behalf by not preparing me***" (See Exhibit #15); and 7) Mr. Gable never told the Court that he was denied his right to testify either in his letter or in open Court, and, in fact, verbalized

3 on the record his concurrence with the Defense's advice that he not testify in the Penalty Phase of the trial.

4

5 c. It is now in retrospect that Mr. Gable has decided that he should have testified because of his conviction as well as his presently reconstructed "Gable Time Line"; i.e., in retrospect Mr. Gable has determined that he could have talked his way out of evidence which was presented against him and

6 avoided conviction. Mr. Gable thus asserts that he was denied his right to testify, which as was previously noted is part and parcel of his Claim that he was denied his "alibi" defense.

7

8 _THIRD CLAIM OF RELIEF_
PETITIONER GABLE REALLEGES SECTIONS 1 THROUGH 8 OF HIS FIRST CLAIM AND

9 ALLEGES THAT HE WAS DENIED HIS RIGHT TO DUE PROCESS UNDER THE FOURTEENTH AMENDMENT TO THE UNITED STATES CONSTITUTION AND HIS RIGHT TO BE HEARD BY

10 HIMSELF UNDER ARTICLE I, SECTIONS 10 AND 11, OF THE OREGON CONSTITUTION IN THAT THE PRESIDING JUDGE, THE HONORABLE C. GREGORY WEST:

11

12 _CLAIM #1:_ FAILED TO GIVE THE DEFENDANT AN OPPORTUNITY TO BE HEARD ON HIS LETTER DATED APRIL 2ND, 1991, IN WHICH HE REQUESTED A HEARING ON HIS ATTORNEYS FAILURE TO CONSULT WITH HIM AND PREPARE FOR TRIAL.

13

14 _FINDINGS and CONCLUSIONS:_

15 a. Contrary to the Claim regarding a letter dated April 2nd, 1991, the letter is actually dated 3/31/91. See Exhibit #14. Contrary to the assertion in the Claim the request in the 3/31/91 letter was not for a hearing but rather "What I am asking is if that you allow a continuance and that somehow you

16 would get my defense team to get involved with me on a consistent level." Mr. Gable did say in conclusion "If it takes a court appearance to put this on record, I would like that done soon as possible."

17 See Pg. 2 of Exhibit #14.

18 b. The Trial Judge simply did what he would normally do with this type of letter – he passed it on to Defense Counsel for action as they might deem appropriate. No denial of due process is involved

19 in the Court handling the letter as it did and not setting the matter down for a Court hearing.

20

21 _CLAIM #2:_ FAILED TO GIVE THE DEFENDANT AN OPPORTUNITY TO BE HEARD ON HIS LETTER DATED JULY 1ST, 1991, IN WHICH HE ASKED THE COURT FOR RELIEF BASED ON SEVERAL MATTERS INCLUDING LACK OF PREPARATION, TAKING AWAY DEFENDANT'S

22 RIGHT TO TESTIFY, AND ODOR ON THE BREATH OF DEFENSE ATTORNEY ABLE.

23 _FINDINGS and CONCLUSIONS:_

24 a. As previously noted, the matter involving the 7/1/91 letter (See Exhibit #15) was briefly addressed by Mr. Gable in Court on the record on July 1st, 1991 (See Exhibit #232, Pg. 10082) and by

25 the Court on July 2nd, 1991 (See Exhibit #232, Pg. 10085). It thus cannot be contended by Mr. Gable that he did not have an opportunity to be heard, other than perhaps that he may have wanted more done

26 than was done about his concerns.

27 b. There was no deprivation of Mr. Gable's rights when the Trial Judge handled the 7/1/91 letter

28 _Frank Edward Gable vs. State of Oregon Post-Conviction Judgment (a:\PCGable121500.JUD)_   Page -98-

1 in the manner it which it was handled.

2

CLAIM #3: FAILED TO POSTPONE THE TRIAL OR GRANT OTHER APPROPRIATE RELIEF
3 WHEN ALMOST ALL OF THE DEFENSE INVESTIGATORS PRESENTED A LETTER TO
ROBERT L. ABEL AND JOHN E. STORKEL, THAT WAS DELIVERED TO THE COURT,
4 INDICATING THAT THE DEFENSE ATTORNEYS WERE NOT PREPARED TO PROCEED TO
TRIAL.

5

FINDINGS and CONCLUSIONS:

6

      a. This Claim has been previously addressed the particulars of the Defense Investigators'
7 perception that more time was needed to prepare and the resulting Motion and Affidavit for Continuance
under the Summary of Mr. Gable's post-conviction case-in-chief and difficulties associated with the
8 Defense Investigators' letter to Judge West under the discussion of other Claims. No proper basis was
established for the requested continuance, and Judge West was correct in his denial of the request.

9

      b. It should also be noted that the Defense Investigators' perception that more time was needed
10 resulting in the letter addressed to Judge West occurred near the end of January of 1991 with a hearing
on the Motion and Affidavit for Continuance heard on 2/4/91. Trial did not begin until 3/4/91 and lasted
11 jury selection took the remainder of March and April of 1991 with the State's Opening Statement not
being delivered until 5/1/91. There thus was a passage of almost three (3) months from the time of their
12 request until the State's opening statement during which the Defense Investigators could pursue the
investigation. In addition, the Guilt Phase of the trial was not concluded until 6/21/00, four and one-half
13 (4 ½) months after the request for continuance. There thus was ample time for the Defense to pursue
whatever needed to be pursued during that passage of time. However, dealing with all of the Defense
14 Investigators' pet theories, or for that matter any one of those pet theories, whatever those theories may
have involved, has not been shown in this case as to have been necessary or something that would have
15 been productive.

16

17 **FOURTH CLAIM OF RELIEF**
      PETITIONER GABLE REALLEGES SECTIONS 1 THROUGH 8 OF HIS FIRST CLAIM AND
18 ALLEGES THAT THE COURT LACKED JURISDICTION OF THE DEFENDANT BECAUSE THE
INDICTMENT ON WHICH HE WAS CHARGED WAS INVALID FOR THE FOLLOWING
19 REASONS:

20 CLAIM #1: THE FOREMAN OF THE GRAND JURY, THOMAS H. DENNEY, OSB NO. 66034,
WAS A CAREER PROSECUTOR EMPLOYED BY THE DEPARTMENT OF JUSTICE OF THE
21 STATE OF OREGON, WHICH WAS INVOLVED IN THE INVESTIGATION OF THE FRANCKE
HOMICIDE AND SHOULD HAVE BEEN EXCUSED FROM SERVICE ON THE GRAND JURY
22 PURSUANT TO ORS 10.050(2).

23 FINDINGS and CONCLUSIONS:

24   a. See Findings and Conclusions under Claim #13 of Mr. Gable's First Claim for relief which
are incorporated under this section.

25

      b. Although the Court could possibly have used its discretion to excuse Thomas H. Denney as
26 a Grand Juror under ORS 10.050(2) [but only by stretching the words "whose presence on the jury
would substantially impair the progress of the action on trial or prejudice the parties thereto", since no
27 substantial impairment or prejudice was shown], in the event it received a proper request to do so, no

28 Frank Edward Gable vs. State of Oregon Post-Conviction Judgment (a:\PCGable121500.JUD)   Page -99-

589

1   request was made to the Court to exercise such discretion. No factual basis was shown which would have provided a basis for excusing Mr. Denney as a Grand Juror, and no legal basis was established or

2   shown as to why Thomas H. Denney should have been removed as a Grand Juror.

3   (1) In any event, even if this was a potential infirmity in the indictment process, it would not result in a void conviction and thus would not be a basis for post-conviction relief. See *Goodwin*

4   *vs. State*, 125 OrApp 359 (1993).

5   *CLAIM #2:* OREGON STATE POLICE OFFICER WILLIAM PIERCE WAS PERMITTED TO SIT IN ON THE GRAND JURY PROCEEDINGS IN VIOLATION OF ORS 132.090.

6

  FINDINGS and CONCLUSIONS:

7

8   a. See Findings and Conclusions under Claim #13 of Mr. Gable's First Claim for relief which are incorporated under this section.

9   b. Under ORS 132.090(2) the Court had authority for entry of the Order allowing Oregon State Police Officer William Pierce to attend the Grand Jury proceedings. In any event, even if there was no

10   such statutory authority, the defect could only have made the Indictment in this case voidable; not void, *ab initio.* See *Goodwin vs. State*, 125 OrApp 359, 361 (1993).

11

12   ***FIFTH CLAIM OF RELIEF***
        PETITIONER GABLE REALLEGES SECTIONS 1 THROUGH 8 OF HIS FIRST CLAIM AND

13   ALLEGES THAT HE WAS DENIED EFFECTIVE ASSISTANCE OF APPELLATE COUNSEL UNDER THE SIXTH AND FOURTEENTH AMENDMENTS TO THE UNITED STATES

14   CONSTITUTION AND UNDER ARTICLE I, SECTION 11, OF THE OREGON STATE CONSTITUTION IN THAT APPELLATE COUNSEL:

15

16   *CLAIM #1:* FAILED TO PROPERLY AND ADEQUATELY ARGUE ALL ISSUES ADEQUATELY RAISED BY TRIAL COUNSEL. IN PARTICULAR, THE TRIAL COURT'S FAILURE TO ALLOW PETITIONER'S ATTORNEYS TO PRESENT EVIDENCE THAT TIMOTHY NATIVIDAD AND/OR

17   JOHN CROUSE WERE INVOLVED IN THE MURDER OF MICHAEL FRANCKE.

18   *FINDINGS and CONCLUSIONS:*

19   a. In the mass of the Exhibits entered as a part of this post-conviction relief process, this Court has been unable to locate materials associated with Mr. Gable's Appeal, which obviously was pursued

20   in his case since Claim is made of ineffective assistance of appellate counsel. This Court has not found any Notice of Appeal, any Delineation of Issues, or any Parties' Briefs in order to know what issues were

21   presented by Appellate Counsel to the Appellate Courts.

22   b. In any event, this Court finds nothing, other than the *ex post facto* matter involving the true life option, which would, in this Court's view, constitute a viable appellate issue which could have led

23   to a potentially prejudicial effect upon or to the detriment of Mr. Gable. The *ex post facto* is not included in the claim of ineffective assistance of Appellate Counsel. This Court has already concluded

24   that the *ex post facto* protections were waived by Mr. Gable in this case, and this Court does conclude that there was no detriment or prejudicial effect on Mr. Gable by not raising such on appeal, in the event

25   that this issue was in fact not raised on appeal.

26   c. The lack of merit with respect to potential appellate issues includes the Timothy Natividad and/or John Crouse matters.

27

28   *Frank Edward Gable vs. State of Oregon Post-Conviction Judgment (a:\PCGable121500.JUD)*   Page -100-

1 CLAIM #2: FAILED TO PROPERLY RAISE ON APPEAL THAT THE INDICTMENT WAS IN VIOLATION OF ORS 10.050(2) BECAUSE THE FOREMAN, THOMAS H. DENNEY, OSB NO.
2 66034 WAS A CAREER PROSECUTOR EMPLOYED BY THE DEPARTMENT OF JUSTICE OF THE STATE OF OREGON AND BECAUSE UNAUTHORIZED PERSONS WERE ALLOWED TO
3 SIT IN DURING GRAND JURY TESTIMONY AND DELIBERATIONS IN VIOLATION OF ORS 132.090.

4

FINDINGS and CONCLUSIONS:

5
        a. There was no viable appellate issue involving Thomas H. Denney sitting as Foreperson of the
6 Grand Jury that indicted Mr. Gable.

7   b. There was no viable appellate issue involving the Order which allowed Oregon State Police
Officer Pierce sit in on Grand Jury proceedings.

8

9 *SIXTH CLAIM OF RELIEF*
        PETITIONER GABLE REALLEGES SECTIONS 1 THROUGH 8 OF HIS FIRST CLAIM AND
10 ALLEGES THAT HE WAS DENIED DUE PROCESS UNDER THE FOURTEENTH AMENDMENT UNDER THE UNITED STATES CONSTITUTION AND HIS RIGHTS UNDER ARTICLE VI AND
11 HIS RIGHTS UNDER ARTICLE I, SECTION 11, OF THE OREGON STATE CONSTITUTION FOR THE FOLLOWING REASONS:

12

13 CLAIM #1: THAT PROSECUTORS BOSTWICK AND MOORE FAILED TO DISCLOSE EXCULPATORY EVIDENCE AND FAILED TO DISCLOSE PLEA AGREEMENTS OR PROMISES
14 TO THE FOLLOWING KEY WITNESSES:

15   Before even beginning a discussion of this Section, it should be noted that Mr. Hadley noted that this Sixth Claim of Relief constituted a non-issue for purposes of these post-conviction proceedings.
16 See PCT, Vol. II, Pg. 329. However, where applicable this Court has nevertheless made findings and conclusions from what was revealed in the trial transcript.

17
        a. JODIE SWEARINGEN
18

FINDINGS and CONCLUSIONS:
19
        Jodie Swearingen was called and testified for the Defense. Exhibit #157 does contain letters
20 dated 1/10/91 and 3/25/91 and copies of attached materials concerning the contacts between the State and Ms. Swearingen including compensation for the period she was held as a material witness.

21
        b. CAPPIE HARDEN
22

FINDINGS and CONCLUSIONS:
23
        During the States' questioning Cappie Harden acknowledged some charges had been dropped
24 against him in plea negotiations, that he had been allowed to enter the drug DROP program, See Exhibit #232, Pgs. 8074- 8075. Such matter was pursued on cross-examination. See Exhibit #232, Pgs.
25 8114 - 8115.
        Exhibit #157 contains a letter sent from the District Attorney's Office to Mr. Abel dated 3/24/91
26 with copies of written materials involving plea agreements with Cappie Harden.

27   c. JANYNE GABLE

28 Frank Edward Gable vs. State of Oregon Post-Conviction Judgment (a:\PCGable121500.JUD)   Page -101-

591

1   *FINDINGS and CONCLUSIONS:*

2   There has been no showing of any plea agreement or consideration accorded to Janyne Gable for her testimony.

3

       d. MIKE KERRINS

4

   FINDINGS and CONCLUSIONS:

5

6   Although Mike Kerrins was not called by the State, there is a letter dated 3/20/91 from the District Attorney's Office to Mr. Able in Exhibit #157 which delineates the contacts and considerations involving Mr. Kerrins.

7

       e. JOHN CROUSE

8

   FINDINGS and CONCLUSIONS:

9

10   John Crouse was not called by the State, and the Defense was prevented from calling Mr. Crouse. The record does not reveal any plea agreement or consideration being accorded to John Crouse by the State of Oregon.

11

       f. JOHN KEVIN WALKER; AND

12

   FINDINGS and CONCLUSIONS:

13

14   During the State's questioning John Kevin Walker testified about what consideration he had received as a result of his testimony which included the provision that there would be no prosecution for any non-person crimes that he revealed through his testimony. See Exhibit #232, Pg. 8183 - 8184.

15   Exhibit #157 also contains four (4) letters dated 4/12/90, 10/2/90, 3/21/91 and , generated by the District Attorney's Office to Mr. Abel and others detailing the nature of considerations and agreements

16 accorded to Mr. Walker.

17   g. KRIS KEERINS

18   *FINDINGS and CONCLUSIONS:*

19   Kris Keerins was not called by either the State or the Defense. The record does not reveal any plea agreement or consideration being accorded to Kris Keerins.

20

21   In addition in an overall sense, the Trial Court itself during discussions regarding discovery matters said during the Defendant's case-in-chief that after a year of involvement with the case that there was no

22 doubt in the Court's mind that Counsel, both the State's or Defendant's, had done nothing to cover up anything relating to the case. See Exhibit #232, Pg. 8641. From this Court's review of the evidence in

23 this case, it is clear that the Trial Judge was correct in his assessment. Everything which should have been disclosed by the State, including plea agreements or considerations accorded to any witness or even

24 potential witness, was disclosed to the Defense.

25

26

27

28   *Frank Edward Gable vs. State of Oregon Post-Conviction Judgment (a:\PCGable121500.JUD)*   Page -102-

ADDITIONAL GENERAL PRINCIPLES OF LAW

1. The standard for judging counsel's performance is essentially identical under both the Oregon and Federal Constitutions. Chew vs. State, 121 OrApp 474, 477, 855 P2d 1120 *rev den* 318 Or 24 (1994).

Under the Oregon Constitution, a petitioner must prove that counsel "failed to exercise professional skill and judgment, failed to diligently and conscientiously advance the defense and that the failure prejudiced his defense"; under the Federal Constitution, a petitioner must prove that "counsel's assistance was unreasonable and "there is a reasonable probability that, but for counsel's unprofessional errors, the result of the proceeding would have been different."

a. "Prejudice" for purposes of post-conviction relief "consists of those acts or omissions 'which would have a tendency to affect the result.'" Stevens vs. State, 32\2 Or 101, 110, 901 P2d 1137 (1995).

2. A defense attorney is not required to follow a client's specific instructions regarding the handling of a client's defense if defense counsel does not feel that such instructions would be in the best interest of the client. Krummacher vs. Gierloff, 290 Or 867, 874, 627 P2d 458 (1981).

"It [*a lawyer's devotion to the interests of the defendant*] does not require that the lawyer must automatically do the defendant's bidding in all respects or otherwise suspend the operation of professional ethics and judgment. Nor does it require that he expend time and energy uselessly or for negligible potential benefit under the circumstances of the case. Rather, it requires that the lawyer do those things reasonably necessary to diligently and conscientiously advance the defense."

3. A Petitioner who seeks post-conviction relief based on an issue not raised at trial must prove that the failure to raise the issue constituted inadequate assistance of trial counsel or otherwise that the petitioner could not have reasonably been expected to raise the issue at trial. Palmer vs. State, 318 Or 352, 867 P2d 1368 (1994).

OVERALL CONCLUSION OF LAW

593

1 Petitioner Gable has not demonstrated that Defense Counsel or Appellate Counsel failed to

2 exercise professional skill or judgment or failed to diligently and conscientiously advance the defense

3 of Frank Gable. Counsels' efforts on Petitioner Gable's behalf were reasonable, and Petitioner Gable

4 has not demonstrated any act or omission which would have had a tendency to affect the result or that

5 would have caused the result of the proceeding to be different than what it was.

6 No prosecutorial error has been shown in these proceedings. No judicial error has been shown

7 in these proceedings, assuming for purposes of argument that matters involving the Trial Court were

8 not appealed or could not have been appealed, which would be required before post-conviction relief

9 could address those matters in any event.

10

11 ### *ORDER AND JUDGMENT*

12 NOW THEREFORE IT IS HEREBY ORDERED THAT JUDGMENT, AS SET FORTH

13 HEREIN, BE, AND IT IS HEREBY, GIVEN IN FAVOR OF DEFENDANT STATE OF OREGON

14 AND AGAINST PETITIONER FRANK EDWARD GABLE IN THESE PROCEEDINGS.

15 DONE AND DATED this _2ND_ day of __JANUARY__, 2001.

16

17 F. J. Yraguen, Senior Circuit Court Judge

18 1/25/2001

19

20 Duplicate Original

21

22

23

24

25

26

27

www.ingramcontent.com/pod-product-compliance
Lightning Source LLC
Chambersburg PA
CBHW031951170526
45157CB00002B/461